Lean Management Solutions for Contemporary Manufacturing Operations

"Applications in the Automotive Industry"

Edited by

Gonzalo F. Taboada

Department of Industrial Engineering,
Universidad Tecnológica Nacional FRC,
Córdoba
Argentina

Lean Management Solutions for Contemporary Manufacturing Operations

"Applications in the Automotive Industry"

Editor: Gonzalo Taboada

ISBN (Online): 978-981-5036-12-1

ISBN (Print): 978-981-5036-13-8

ISBN (Paperback): 978-981-5036-14-5

need for a court order if at any point you breach any terms of this License Agreement. In no event will any delay or failure by Bentham Science Publishers in enforcing your compliance with this License Agreement constitute a waiver of any of its rights.

3. You acknowledge that you have read this License Agreement, and agree to be bound by its terms and conditions. To the extent that any other terms and conditions presented on any website of Bentham Science Publishers conflict with, or are inconsistent with, the terms and conditions set out in this License Agreement, you acknowledge that the terms and conditions set out in this License Agreement shall prevail.

Bentham Science Publishers Pte. Ltd.
80 Robinson Road #02-00
Singapore 068898
Singapore
Email: subscriptions@benthamscience.net

BENTHAM
SCIENCE

CONTENTS

FOREWORD

People are impacted by Science every day. This simple fact is advantageous when it represents an improvement in the quality of life. Many times, when people refer to their quality of life, they associate it with free time. However, this book sheds light on improving the quality of life from the work perspective. Since we probably spend a third of our time working, it is especially important to embrace Science and Technology to produce meaningful changes that benefit us.

These changes, however small they may seem, can have major impacts on the society in which they are generated. Incorporating Science and Technology in everyday work-life can help increase opportunities, improve mental well-being, and enhance productivity which are central to a better quality of life. These changes could substantially contribute towards the incorporation of people with special abilities into today's industry.

Dr. Juan Pablo Zanin
Rutgers University, New Jersey, EEUU
USA

PREFACE

Lean Management Solutions for Contemporary Manufacturing Operations opens an illustration of some of the smart functionalities that can be implemented. The book also introduces technologies, soft and hard skills, suitable for the general integration. Following this, the authors provide a comprehensive review of the recent advances in technologies and methods, in different areas, including synergies with human capital.

The following chapters establishes a standard for application of mathematical concepts in workplaces, agile methods for projects, applicated design, low cost solutions. It also includes people development, planning, motivation and especially deals with persons with disabilities in workplace context.

The goal of the work is to open the eyes to the new industry revolution.

Various setups are thoroughly examined by the authors, considering the contribution of each case.

Chapter 1: This article tries to use Euler's concept to solve the problem of the 7 Königsberg bridges, applied in the workplace organization.

This new application of Graph Theory is intended to help improve different jobs, either from the design of new workstations, as well as opportunities for improvement for those workstations current.

Chapter 2: This chapter explores the concept of non-value-added operations in the manufacturing industry and how it impacts part design. The idea is to describe what are currently considered operations that do not add value to manufacturing and show examples of applied design, giving an idea of how the designer or engineer should think when making a new product, they should not only focus in functionality, also in the process and always contextualizing the different technologies available, current laws and the market to which it is trying to introduce.

Chapter 3: In times where the fourth industrial revolution is looming on the near horizon, technological advances and new employment configurations invite us to question access to quality jobs for vulnerable groups such as people with disabilities. This requires investigating the management of resources, physical, technical and procedural factors involved in the design of jobs and reviewing some alternatives such as the Supported Employment methodology.

Chapter 4: The influence of strategic planning can be a link or obstacle to the good performance of the company and can be a factor of distinction and influence in the commitment of those who are part of it. Strategic planning must be understood as a participatory process, which will not solve all uncertainties, but which must draw a line of those affected to act accordingly.

Chapter 5: In this article we are going to review different solutions for material handling problems to increase productivity.

In this process we are learning some tools used to analyze time and methods and improve them.

The aim is to demonstrate that automation is not always expensive, if we use our brain we can find out mechanism cheaper than a robot.

Chapter 6: This article consists of analyzing the feasibility of using agile methodologies tools in industrial manufacturing projects.

Since its inception, agile management tools have been used for software development projects and technology innovation. Currently, manufacturing projects use only traditional project management methodologies; the challenge is to apply agile tools in traditional management.

Gonzalo F. Taboada
Department of Mechanical Engineering
Universidad Tecnológica Nacional FRC
Córdoba
Argentina

List of Contributors

Cristian F. Stiefkens — Industrial Compass, Córdoba, Argentina

Gerardo E. Rodríguez — Department of Industrial Engineering, Universidad Tecnológica Nacional FRC, Córdoba, Argentina

Gonzalo F. Taboada — Department of Mechanical Engineering, Universidad Tecnológica Nacional FRC, Córdoba, Argentina

Gustavo R. Mena — Department of Psychology, Universidad Nacional de la Rioja, La Rioja, Argentina

Natalia Nissen — Dept. of Legal Medicine, Psychiatry and Pathology, Universidad Complutense, Madrid, Spain

CHAPTER 1

Application of Graph Theory in Workplace Design

Gonzalo F. Taboada[1,*]

[1] *Department of Mechanical Engineering, Universidad Tecnológica Nacional FRC., Córdoba, Argentina*

Abstract: This article tries to use Euler's concept to solve the problem of the 7 Königsberg bridges, applied in the workplace organization.

First of all, it is important to keep in mind that Euler was a mathematician who in 1736 proposed a solution to a problem posed at the time that consisted in taking a walk through the city of Kaliningrad, starting from one of its regions, crossing once all its bridges over the Pregolya River and returning to the same region from which he had started.

Keywords: Assembly, Bridges, Chart, CILR, Design, Diagram, Euler, Eulerian Cycle, Eulerian Way, Graph Theory, Movements, Regions, Route, Workplace, Workstations.

INTRODUCTION

Before entering fully into the article, we will briefly review Leonhard Euler (1707 - 1783) Swiss mathematician (Fig.**1**), main promoter of mathematical analysis in the 18th century, with more than 50 books published, on mathematical analysis, algebra, fluid mechanics, astronomy, *etc.*

Fig. (1). Leonhard Paul Euler.

* **Corresponding author Gonzalo F. Taboada:** Department of Mechanical Engineering, Universidad Tecnológica Nacional FRC., Córdoba, Argentina; Tel: +54 9 351 2412085; E-mail: gonzalo.f.taboada@gmail.com

Amongst his many discoveries and developments, Euler is credited for introducing the Greek letter π to denominate the Archimedes constant (the ratio of a circle's circumference to its diameter), and for developing a new mathematical constant, the "e" (also known as Euler's Number), which is equivalent to a logarithm's natural base, and has several applications such as to calculate compound interest.

Another application that we will develop is the Graph Theory.

Graph Theory is a branch of mathematics that studies flow and/or movements through networks of points and lines by means of graphical representation.

The subject of graph theory had its beginnings in recreational mathematical problems (see numbers game), but has become an important area of mathematical research, with applications in chemistry, social sciences, and computer science [1].

PROPOSED DESIGN

Seven Königsberg Bridges

As we mentioned in the beginning, Euler tried to solve the problem posed on the crossing of the 7 bridges (Fig.2).

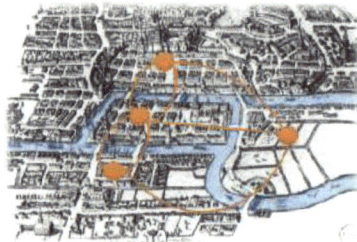

Fig. (2). Seven Königsberg Bridges.

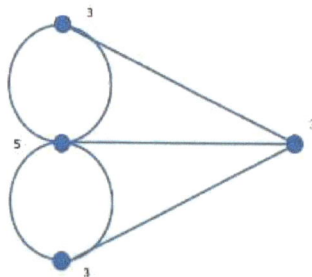

Fig. (3). Graph.

The problem posed has no solution, it is impossible to get a route that meets the condition of connecting all regions using all bridges only one time.

From this problem, Euler enunciated the Graph Theory, which we will use in this chapter to achieve greater efficiency and organization in the different workstations, that is, we will show one of the possible applications of Euler's Theory [2, 3].

Graph Theory

It is the graphical and simplified representation made by Euler to try to solve the problem of the 7 bridges (Fig.3), based on what was called position geometry. Euler designed a graph that represents regions as nodes or vertices and routes (in this case bridges) as edges of that representation.

For a graph to be solved, with the premise of finding a route that crosses all bridges only once, two necessary and sufficient conditions must be met [4].

1. If an even number of bridges arrive at each of the regions of the graph, then the path starts and ends at the same place. This is an Eulerian Cycle (Fig.4).

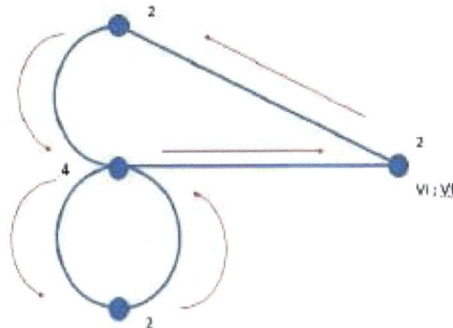

Fig. (4). Eulerian Cycle.

2. If there is a region where an odd number of bridges arrive, then there are exactly two regions with that configuration and the trajectory starts in one of them and ends in the other, where an odd number of bridges also arrive. This is called an Eulerian Way (Fig.5).

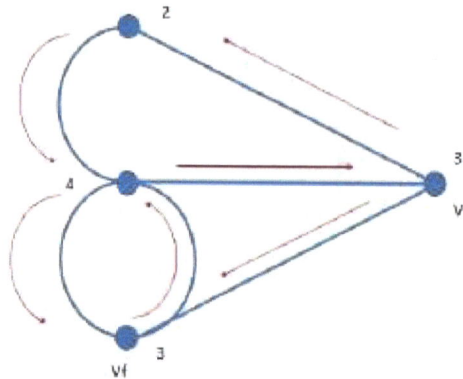

Fig. (5). Eulerian Way.

If the Graph configuration contains more than two nodes where the number of bridges arriving is odd, we can say that the route has no solution for a Cyclic or Eulerian Way.

The application of Graph Theory is wide, it is used for example for:

- Computer Network Design
- Programming and Distribution of Public Services.
- Urban Planning
- Mathematical Modeling
- Cartography

In this paper we will list a new application, which will be the "Design and Organization of the Workplace".

CASE STUDIES

Example 1:

In this case, for example, the assembly of the wheels of a commercial vehicle (Fig.**6**).

Next is the analysis of the current operation (Fig.**7**). It is possible to use graph theory to design a new operation more efficiently.

The route shown in the current operation is inefficient, since for the assembly of the rear wheels, the same route or bridge is traversed three times.

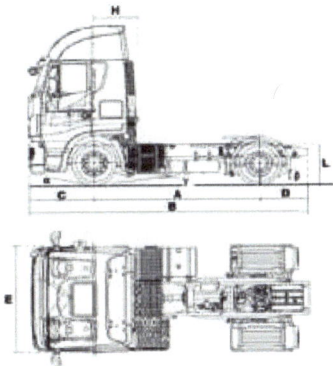

Fig. (6). Assembly Line.

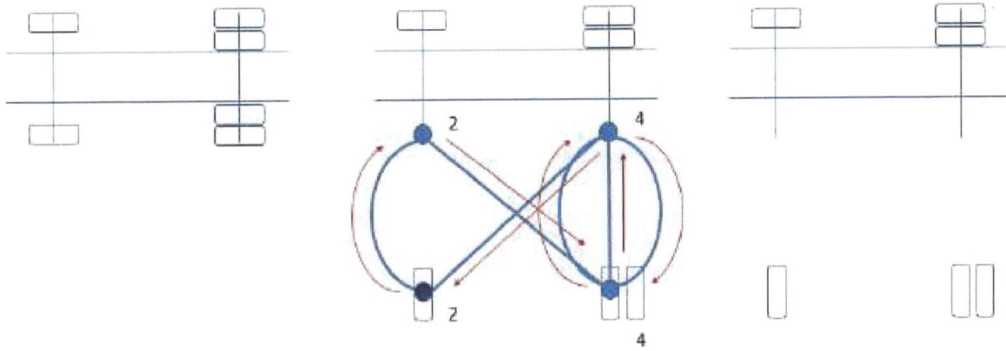

Fig. (7). Current Assembly.

The proposed improvement consists of modifying the mounting device of the dual wheels so that they can be mounted in pairs and thus the proposed path becomes an Eulerian Cycle (Fig.**8**).

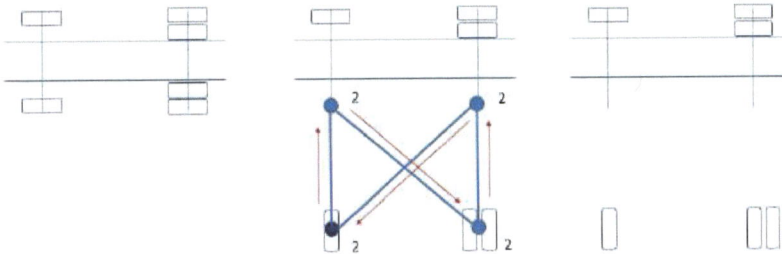

Fig. (8). Eulerian Cycle.

Obviously, there is a noticeable change in the number of operations for the assembly of the rear wheels, from three movements to one, then the cost/benefit ratio calculation must be performed to obtain the real savings of the operation.

Next, we first analyze the path that serves an Eulerian Cycle or Eulerian Way (which will be the most efficient path). To do this we perform the same process that Euler will use in the 7 bridges problem. Each point from which we start and to which we arrive we will call it Regions and we will identify it with letters, and we will enumerate the routes, which will be called Bridges.

If we have one front wheel and two rear wheels to mount from two different positions, we can identify four ABCD regions (Fig.9).

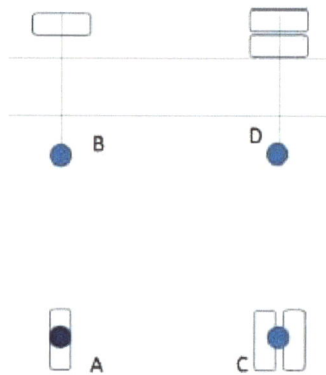

Fig. (9). Regions.

If the bridges to cover the 4 regions are 1, 2, 3, 4; we can matrix the scenarios and determine if the Graph Theory is fulfilled.

The route of the first case presented ABBCCDDCCDDA does not comply with the Graph Theory, since it does not pass only once through the bridge 3 located between the CD segments.

The second proposed route ABBCCDDA represents an Eulerian Cycle according to the graph theorem proposed by Euler and is therefore the ideal route (Fig.10).

As an example, if we consider that the distance of each segment is equal to 1 step, (the step of an average person is equal to 0.8 meters) we will multiply by the number of segments of each of the routes and objectively observe the proposed improvements.

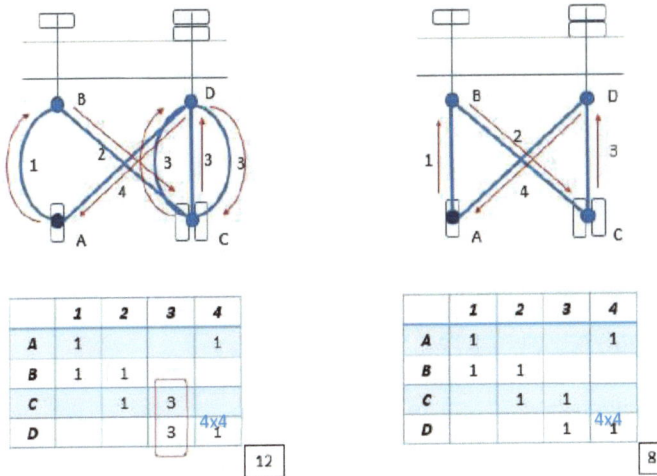

	1	2	3	4
A	1			1
B	1	1		
C		1	3	
D			3	4x4 1

12

	1	2	3	4
A	1			1
B	1	1		
C		1	1	
D			1	4x4 1

8

Fig. (10). Graph Analysis.

Current Assembly Eulerian Cycle

ABBCCDDCCDDA = 6 steps ABBCCDDA = 4 steps

6 x 0,8 = 4,8 m 4 x 0,8 = 3,2 m

Efficiency 33%

On the other hand, it is possible to make a direct comparison between two proposed routes and see the efficiency in a dimensionless number.

Example 2:

We will now show another example of the new application of Graph Theory, this time going from the simple to the complex, moving from workstation design to part design to improve efficiency.

The operation we are going to analyze is the assembly of the banjo and banjo bolt on a fuel filter (Figs.**11** and **12**). This operation consists of placing 4 parts on the filter, 2 washers, a banjo and a banjo bolt.

First, we will analyze the standard assembly and then each of the possible improvements, always from an optimal position (golden zone) and using both hands, *i.e.* an optimal assembly from an ergonomic point of view.

Fig. (11). Materials in the Workplace.

Fig. (12). Fuel Filter.

Standard Assembly: It consists of taking a banjo bolt with one hand and a washer with the other, then both are put together, once the first washer is inserted in the banjo bolt the banjo is taken with the left hand and inserted in the banjo bolt, then with the same hand the last washer is taken and placed in the banjo bolt, finally the set is taken with the right hand and positioned in the filter body and with the left hand the tool is taken and then the corresponding torque is given (Fig.**13**).

(right hand) + (left hand)

ABBFFGGA + CDDFFEEFFDDFFCCGGC = 13

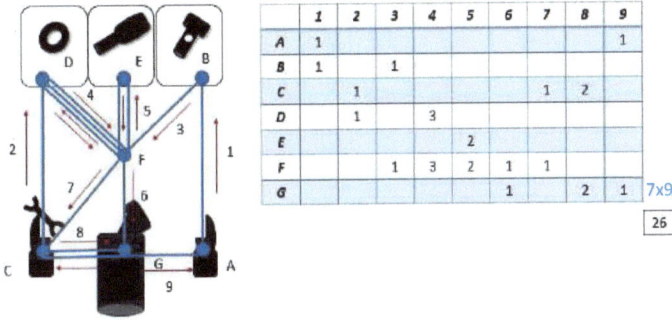

	1	2	3	4	5	6	7	8	9
A	1								1
B	1		1						
C		1					1	2	
D		1		3					
E						2			
F			1	3	2	1	1		
G						1		2	1

7x9

26

Fig. (13). Standard Assembly.

Standard Assembly with Separate Washers: In this second example washers were placed in two opposite sectors to make an improvement in the amount of movement, only that the operator must improve his skills because with the right hand first takes a washer and inserts it into a banjo bolt with the same hand, on the contrary with the left hand takes a washer and with the same hand takes the banjo, then both hands assemble the assembly and then with the right hand is positioned in the body of the filter, the left hand takes the tool to give the corresponding torque (Fig.**14**).

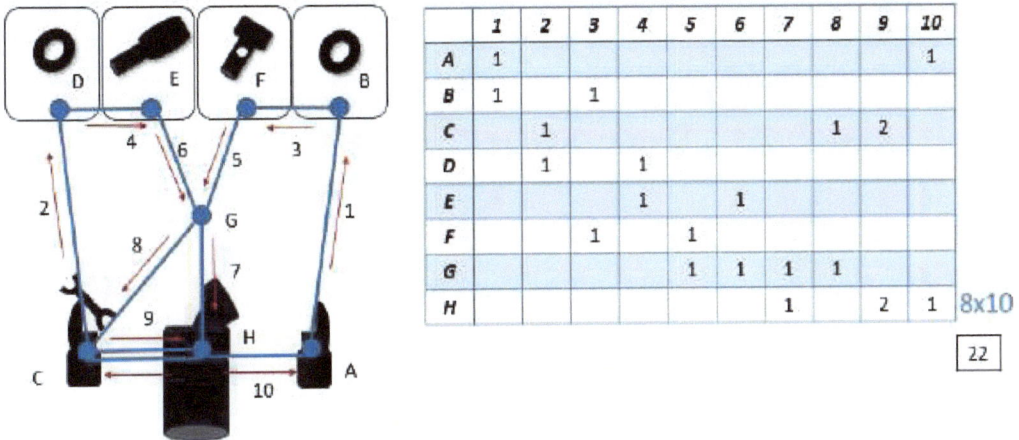

	1	2	3	4	5	6	7	8	9	10
A	1									1
B	1		1							
C		1						1	2	
D		1		1						
E					1		1			
F			1		1					
G						1	1	1	1	
H							1		2	1

8x10

22

Fig. (14). Standard Assembly with Separate Washers.

(right hand) + (left hand)

ABBFFGGHHA + CDDEEGGCCHHC = 11

Double Washer Mounting: In this third case an improvement in product design is used, which will be discussed in more detail in later chapters, double washers are used to reduce movement at a reasonably higher cost per part. First with the right hand the banjo is taken and with the left hand the double washer is taken, they are joined together and then the banjo bolt is taken and inserted assembling the assembly, then with the right hand it is placed on the filter body and the left hand searches for the tool and then applies the correct torque (Fig.15).

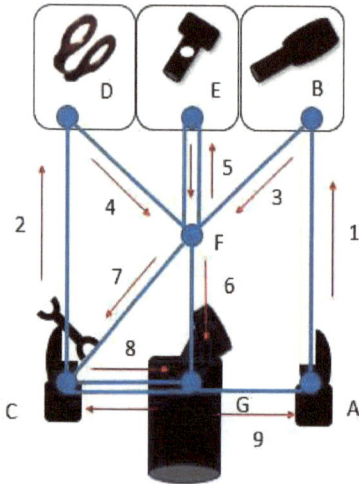

	1	2	3	4	5	6	7	8	9
A	1								1
B	1		1						
C		1					1	2	
D		1		1					
E					2				
F			1	1	2	1	1		
G						1		2	1

7x9

22

Fig. (15). Double Washer Mounting.

(right hand) + (left hand)

ABBFFGGA + CDDFFEEFFCCGGC = 11

Assembly with Banjo WOW System [©]: This case is like the previous one in which an improvement is made in the design of the pieces, the banjo used is an own development in which washers are not used, and the piece is lined with a polymer to perform the sealing function. With this system we reduce not only the movements but also the parts involved, first we take a banjo bolt with the right hand and with the left hand the banjo, then we insert one in the other and then we place it in the body of the filter and the left hand looks for the tool and then applies the corresponding torque (Fig.16).

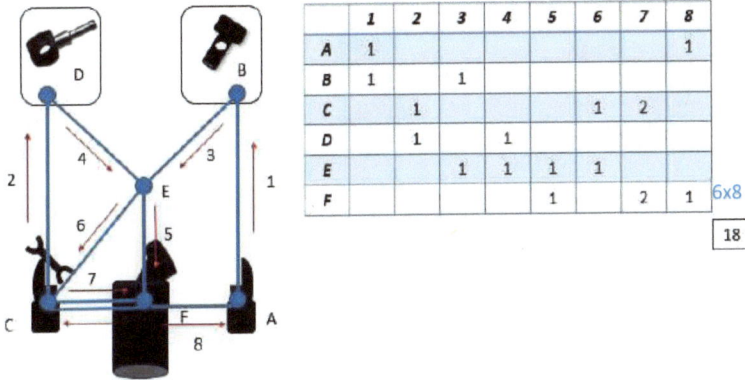

	1	2	3	4	5	6	7	8
A	1							1
B	1		1					
C		1				1	2	
D		1		1				
E			1	1	1	1		
F						1	2	1

6x8

18

Fig. (16). Montaje con Banjo WOW System©.

(right hand) + (left hand)

ABBEEFFA + CDDEECCFFC = 9

Example 3

Another interesting application of graph theory arises when it is linked to the concepts of Autonomous Maintenance.

The main objective of autonomous maintenance is to use the equipment optimally (Fig.17).

Equipment is designed to be reliable, but is often susceptible to breakdowns, defects and minor shutdowns due to lack of basic conditions.

To restore basic machine conditions, autonomous maintenance works as a systematic approach:

- Applying standards of Cleaning, Inspection, Lubrication and refastening (CILR).
- Eliminating dirt sources and difficult access areas with application of machine continuous improvement.

All these activities are carried out directly by the machine operators, who avoid deterioration of the machines and increase their skills according to the evolution of the activity.

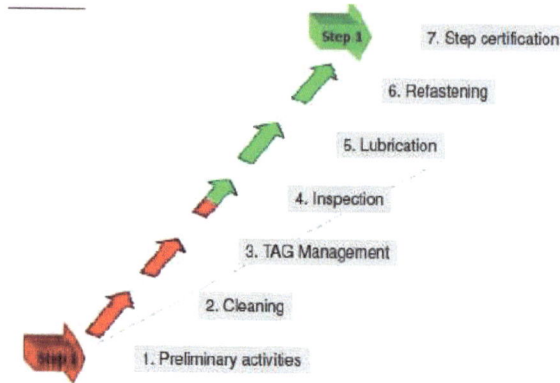

Fig. (17). Initial cleaning and inspection

The way to introduce the method is to design the most efficient route for cleaning, inspection, lubrication, and re-greasing.

This means that the time required for these tasks and, therefore, their costs are reduced (Fig.**18**).

Fig. (18). Device Cleaning without Graph Theory.

By standardizing the traced route, we achieve, firstly, an improvement in cleaning and inspection times, as we have already mentioned, but also ensure that all workers perform the same task regardless of their skills, since each point has specifically defined elements (Fig.**19**).

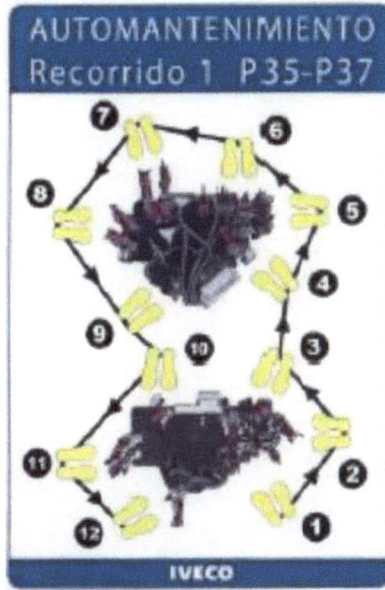

Fig. (19). Device Cleaning with Graph Theory.

The way to improve the routes drawn according to an Eulerian Cycle is to work on the reduction of the segments (Fig.**20**).

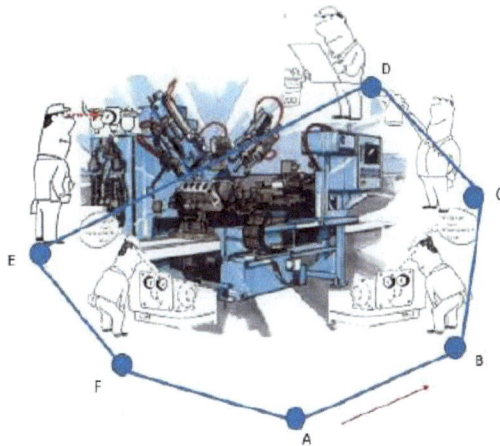

Fig. (20). CILR Route.

For example, in the case of pressure gauge inspection, they can be repositioned so that they can be checked at a single inspection point to make the task more efficient (Figs **21** and **22**).

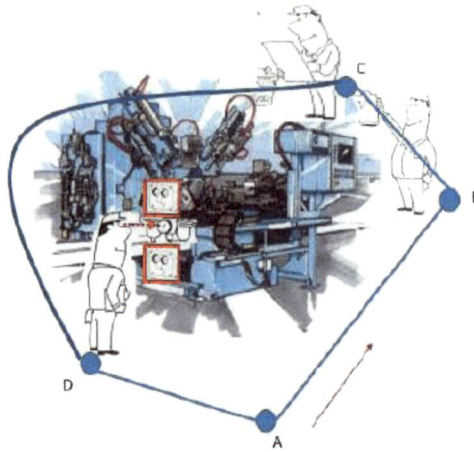

Fig. (21). Eulerian cycle improvement in CILR.

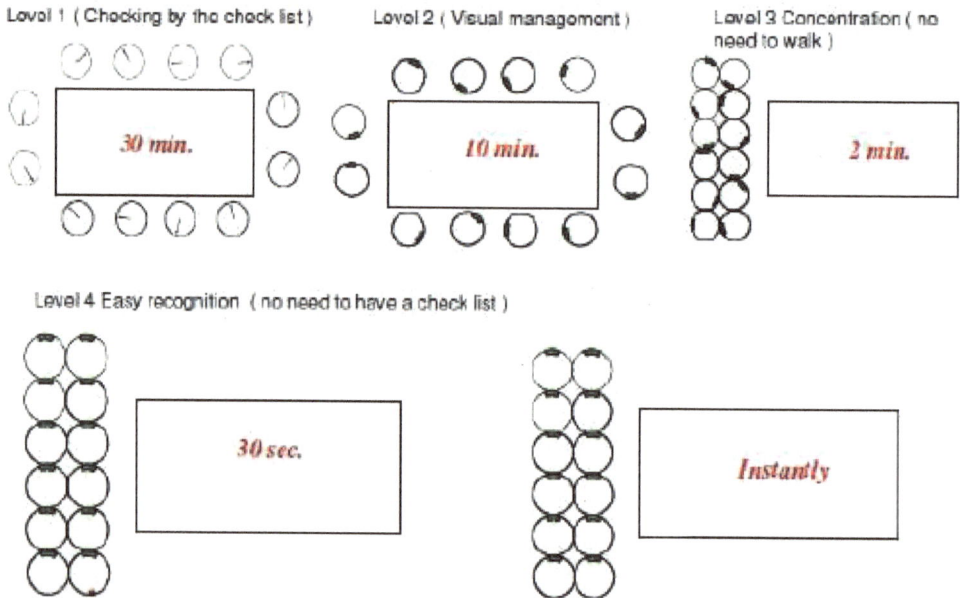

Fig. (22). Improved gauge arrangement.

It follows that the larger the matrix, the lower the efficiency, either because there are more movements (bridges) or more regions.

This new application of Graph Theory can be used in conjunction with or as a replacement for tools such as the "Process Flow Diagram" and the "Spaghetti Chart".

The Process Flow Diagram is represented by a sheet with a format where the specific data of the product or material is detailed, where there are 5 standard symbols to indicate a series of operations, this diagram is very valid to record hidden costs such as distances traveled, delays and temporary storage (Fig.23). The process to be controlled is enlisted in a series of operations that compose it and then the different symbols are joined with lines and the distances and times are added up, to then analyze what is the best solution to make a more efficient process [5].

OPERATION	PROCESS FLOW DIAGRAM					
● o.i. scrawing	LOCATION:			SUMMARY		
	ACTIVITY:			ACTIVITY	ACTUAL	PROPOSED	SAVING
TRANSPORTATION ➡ e.i. material movement	DATE:			OPERATION			
	OPERATOR:			TRANSPORTATION			
	SUPERVISOR:			STORAGE			
	METHOD:			DELAY			
STORAGE ▼ o.i. transfer of material to the warehouse	ACTUAL		PROPOSED	INSPECTION			
	TYPE:			TIME			
	MANPOWER	MATERIAL	MACHINE	DISTANCE			
DELAY D o.i. material in process	COMMENTS:			COST			
INSPECTION ■ e.i. quality control	DESCRIPTION	SYMBOL	TIME	DISTANCE	RECOMMENDATION
		O ➡ ▽ D □			
		O ➡ ▽ D □			
		O ➡ ▽ D □			
		O ➡ ▽ D □			

Fig. (23). Symbol & Process Flow Diagram.

The Spaghetti Chart or Spaghetti Diagram is a way to graphically represent the movements of people, it can be done manually or automatically if the individual is connected to a telemetry device and a suitable software translates the movements in the distance of each one of them, then they are analyzed to improve the layout of machines, parts, *etc.* and thus achieve more efficient operations (Fig.24). To do this, first draw the work area and then, right there, draw the paths taken (if you count steps instead of meters, you should take as a rule that one step equals 0.8m).

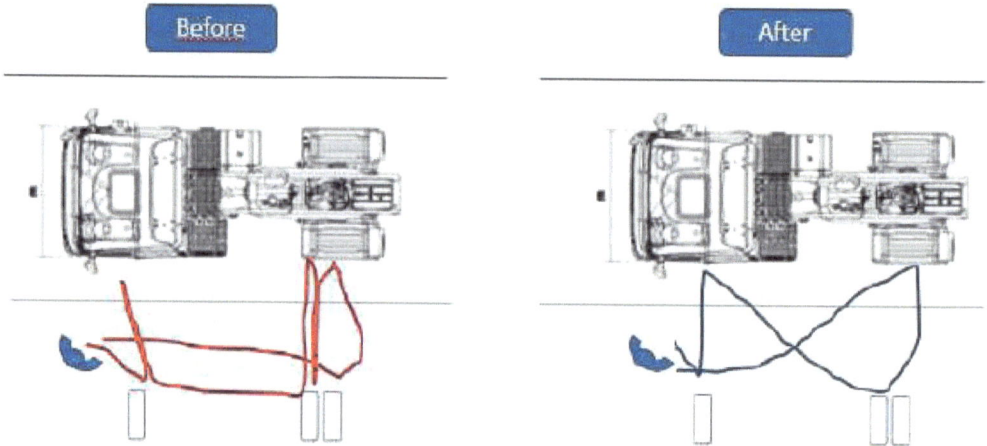

Fig. (24). Spaghetti Chart.

CONCLUDING REMARKS

This new application of Graph Theory is intended to help improve the different jobs, either from the design of new workstations, as well as opportunities for improvement for those workstations current.

The procedure is simple to apply, the workstation is analyzed, the regions and bridges involved in the operation are placed in a matrix, the most efficient operations will be those in which the matrix and its number of movements are smaller, *i.e.*, we can say that:

If matrix A is larger than matrix B, matrix B will be more efficient.

If $|A| > |B| => |B| = E_{(ff)}$

If a matrix A is equal to a matrix B, the one with fewer roads will be the most efficient, *i.e.*, B will be more efficient if it has fewer roads.

If $|A| = |B| \land |B| < ABBn => |B| = E_{(ff)}$

CONSENT FOR PUBLICATION

Not applicable.

CONFLICT OF INTEREST

The author declares no conflict of interest, financial or otherwise.

ACKNOWLEDGEMENTS

Declared none.

REFERENCES

[1] E.R. Aznar. Biografías – Matemático Leonhard Euler.Granada, Spain: Departamento Algebra Universidad de Granada, 2007.

[2] I. Bronshtein, K. Semendiaev. Manual de Matemáticas para Ingenieros y estudiantes - 2da ediciónMoscu, Rusia: Mir 1973.

[3] F.M. Biosca. Enciclopedia Labor Tomo VI – El Lenguaje y Las Matemáticas 2da edición.Barcelona, Spain: Labor S.A., 1962.

[4] N.L. Biggs, E.K. Lloyd, R.J. Wilson. Graph Theory 1736-1936 second edition.Oxford, UK: Oxford University Press, 1986.

[5] B.W. Niebel, A. Freivalds. Methods, Standards and Work Design 11^{a} Ed. DF, Mexico: Alfaomega, 2004.

CHAPTER 2

Applied Design – Efficiency & Added Value

Gonzalo F. Taboada[1,*]

[1] *Department of Mechanical Engineering, Universidad Tecnológica Nacional FRC., Córdoba, Argentina*

Abstract: This chapter explores the concept of non-value-added operations in manufacturing (NVAA; non-value-added-activities) and how it affects part design. In the past, parts were designed to be functional to the mission they would serve, usually without regard to ergonomics, assembly method, and efficiency, as in the aftermarket. The idea is to describe what currently are considered operations that do not add value to manufacturing and to show examples of applied design, giving an idea of how the designer or engineer should think when manufacturing a new product. They should not only focus on functionality but also on the process, always contextualizing the different technologies available, the laws in force and the market to which it is intended to be introduced.

Keywords: Activities, Aftermarket, Applied Design, Banjo, Bolt, Efficiency, Manufacturing, NVAA, Product, Tool, Value, Washer.

INTRODUCTION

Non-value-adding activities (NVAA) for efficient manufacturing pay special attention to the efficiency of different jobs, since they try to minimize or eliminate everything that does not provide added value to the final product. To this end, we will define below the operations that do and do not add value in the manufacturing industry; in order to have a clear idea of where to focus our efforts in the different assembly lines and processes.

After the definition, we will give way to the analysis of how to perform when designing new parts, with a concept of efficiency and functionality of the finished product as well as in the assembly process and its usefulness life. The latter refers to after-sales and of course to the accessibility of the components and their standardization, since the quality of a product is defined by how it performs for the purpose for which it was designed, with the lowest cost and highest efficiency (Fig. **1**).

[*] **Corresponding author Gonzalo F. Taboada:** Department of Mechanical Engineering, Universidad Tecnológica Nacional FRC., Córdoba, Argentina; Tell: +54 9 351 2412085; E-mail: gonzalo.f.taboada@gmail.com

Fig. (1). Operations.

VALUE ADDED OPERATIONS

The operations considered for manufacturing with added value are all those that are assumed to modify the product or transform the product from an initial state to a new one, either a final or partial state; for example, two pieces must be joined by a screw, the adjustment of that screw generates a transformation of our product, they cease to be only two pieces, to become a finished or semi-finished assemble.

NON-VALUE-ADDED OPERATIONS

The operations that do not add value are all those that refer to the contour of the operations that modify the product, that is, operations or micro-operations must be performed so that those that add value can be completed, for example, if a screw must be adjusted, it must first be taken from the shelf and the latter, although it is a necessary operation, does not add any value to the product since it does not transform it. What need to be analyzed are the positions where the screw is to be picked up. The position is not only analyzed as distance to reduce movements with which we reduce time and as we know time translates into an economic account, but we must also analyze the position from the ergonomic point of view since an inadequate position can bring us consequences of physical fatigue or injuries, with which first we will have an injured worker and then we will have associated losses due to the fact that the most skilled operator will not be able to perform the operations. For practical reasons we will take semi-value-added operations as non-value-added operations [1, 2].

PARTS DESIGN

For lean manufacturing much has been written about "Design of Workplaces, their Environment, Tools and Equipment" but most of the time these advances are made as technological solutions for manufacturing and not for product design. This article aims to provide a vision from the applied design or as other authors have called design for manufacturing and assembly (DFMA), which is the sum of design for manufacturing (DFM) and design for assembly (DFA), this means that the design is thought of with a deep knowledge of the technologies and manufacturing means for the plants where they are installed, that is, always taking into account the environmental conditions. However, we must not lose sight of the standardizations since the standards allow us to be more efficient, in terms of manufacturing costs, assembly, storage space, *etc*. I would also like to add a concept which is design thinking about aftermarket (DFAf), since the conservation of components and their replacements must be taken care of from the design conception. Therefore, we will call "applied design" as "design for manufacturing, assembly and aftermarket" (DFMAAf).

Engineering judgment is therefore an important factor in the design and manufacture of reliable components. A typical design project will require detailed information, a good level of understanding and clear decision making on the following aspects.

- Material behavior.
- Product behavior.
- Material and product performance.
- Details of the service loads, working environment.
- Potential failures modes.
- Cost basis of raw materials, manufacturing processes and assembly.
- Design life, maintenance possibilities and consequences of failures.

The following is a series of examples to clarify the concepts explained above, following the methodology of the aspects mentioned above [3].

Hold and Drive

It is an efficient nut tightening system, designed to reduce assembly times in the automotive industry (Figs **2-9**).

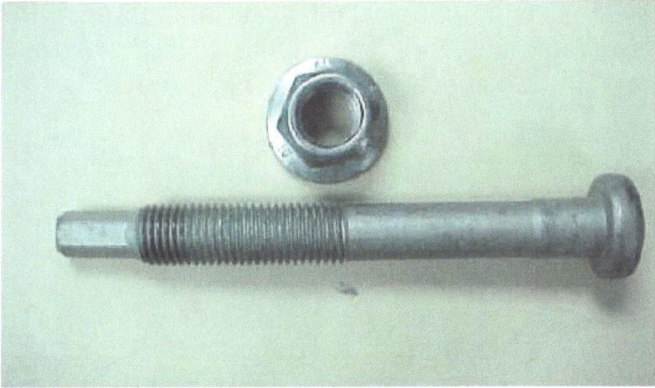

Fig. (2). Hold & Drive Bolt.

We say that it is a system since not only the fasteners (bolts) were redesigned but also the tools (hold & drive nut-runner) that would be used to make the adjustments in the different factories, here we are in the presence of the design for manufacturing and assembly (DFMA).

This system consists of modifying the end of the bolt so that we have an area where we can retain the bolt while adjusting the nut, *i.e.*, what was previously done with a tool (nut tightener or screwdriver) in one hand and another tool in the other to retain the bolt, is now done with a single tool, since the head of this is also specific to be able to retain from the end of the modified bolt and adjust the nut at the same time (Fig. **3**).

Fig. (3). Hold & Drive Application.

The only point that perhaps in the example needs to be addressed is the design with respect to the aftermarket, as many of these new bolts do not have a hex head, which is not a suitable option, perhaps a minimal investment should be made in bolt manufacturing and applied design (DFMAAf). Compared to traditional fasteners and traditional assembly (Figs. **4** and **5**).

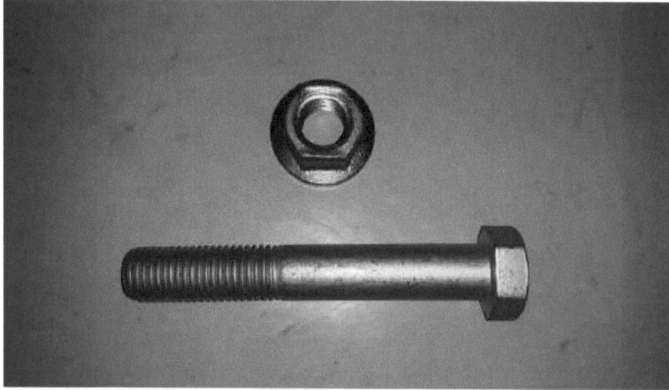

Fig. (4). Traditional Fastener.

Fig. (5). Traditional Assembly.

Let's think that not all dealerships and/or authorized services will have the tools (hold & drive nut-runner).

As mentioned above, the applied design must consider manufacturing, assembly and after-sales efficiency.

Another point within the traditional designs or assemblies and whenever the dimensions and costs allow it, is to design the parts so that the operator can adjust them with only one hand and with the least amount of parts involved, an example within the assembly of shaft elastics can be threaded parts without using nuts (Fig. **6**).

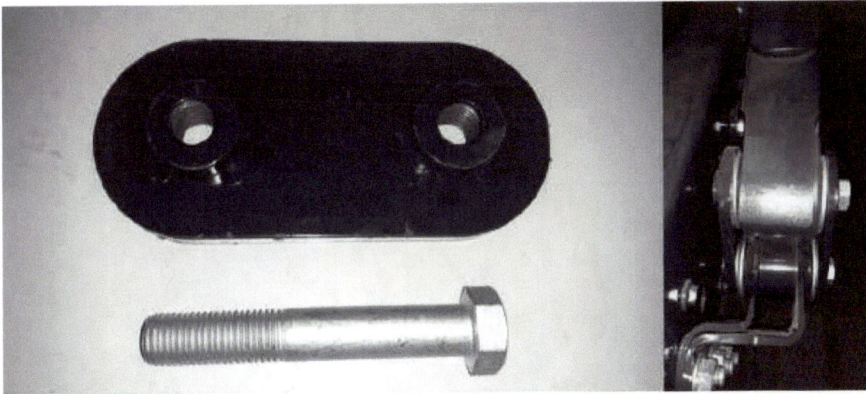

Fig. (6). Other Design.

On the other hand, the assembly method and sequence must also be taken into account when designing, either from an initial design, how and in which part of the assembly line the assembly will be performed and the desired configuration.

We will now show another set of examples of adjustments and emphasize in the assembly sequence.

In the figures above you can see the improvement in the design, when using hold & drive the other hand is free (Fig. **7**).

Fig. (7). The same assembly, with different parts and tools.

Part of the design must also consider the assembly sequence, as the contour of the area surrounding the fitting must allow the tool entry (Fig. **8**).

Fig. (8). Assembly Interference.

Fig. (9). Different, Hold & Drive.

Banjo Fitting

The swivel fitting assembly, also called banjo assembly, has been a very suitable mechanical solution for fluid transmission, which has not undergone major variations over time. The fixing consists of interposing between the banjo and the banjo screw two washers, which are made of a soft material in relation to the surfaces where they are placed to make the seal and prevent fluid leakage.

Advances have occurred mainly in washers and/or seals, making them more efficient and in those cases, we will give some examples of applied design.

Initially, the rectangular section washers were made of copper to ensure sealing. The points to consider are the cost of the material and the way to handle the washers, since if they are bent the copper hardens and does not seal the surfaces well (Fig. **10**).

Fig. (10). Different Assembly with Copper Washer – KTM Motorcycles & CASE Tractors.

Aluminum was added to this type of washer, which is cheaper to obtain, and performs relatively the same functions (Fig. **11**).

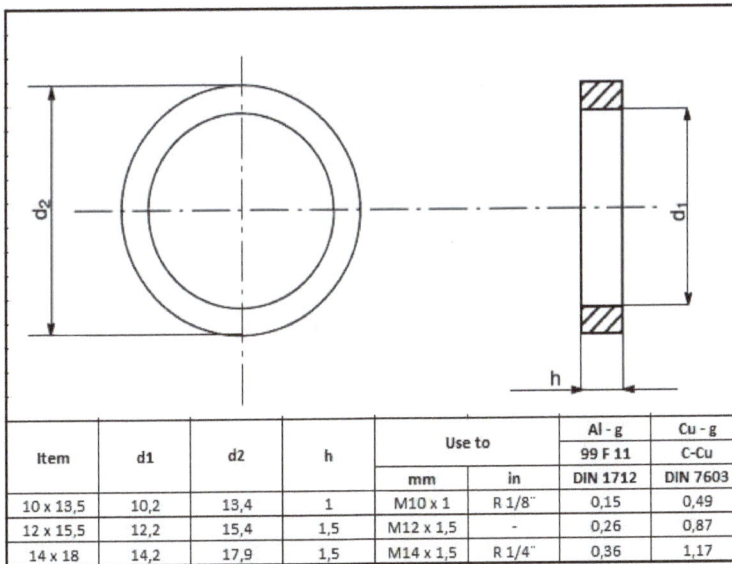

Item	d1	d2	h	Use to		Al - g 99 F 11 DIN 1712	Cu - g C-Cu DIN 7603
				mm	in		
10 x 13,5	10,2	13,4	1	M10 x 1	R 1/8"	0,15	0,49
12 x 15,5	12,2	15,4	1,5	M12 x 1,5	-	0,26	0,87
14 x 18	14,2	17,9	1,5	M14 x 1,5	R 1/4"	0,36	1,17

Fig. (11). Washer, Copper – Aluminum Alloy.

Mixed washers (steel and rubber) also emerged with similar uses, but always taking into account the type of fluid to be sealed, since rubber can be affected (Figs. **12** and **13**).

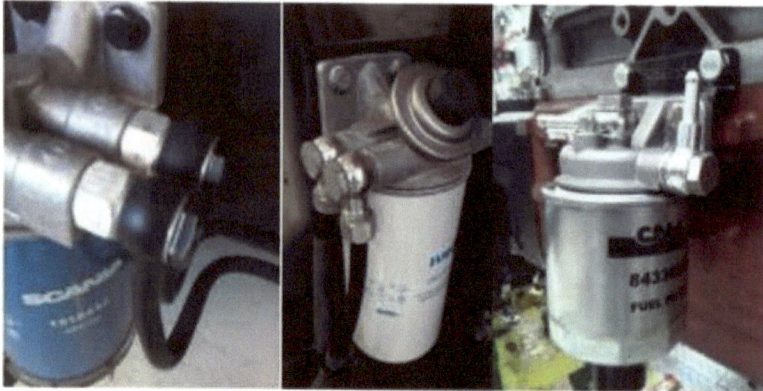

Fig. (12). Fuel Connection with, Aluminum Washer (SCANIA & IVECO Trucks) & Steel Rubber Washer (New Holland Tractors).

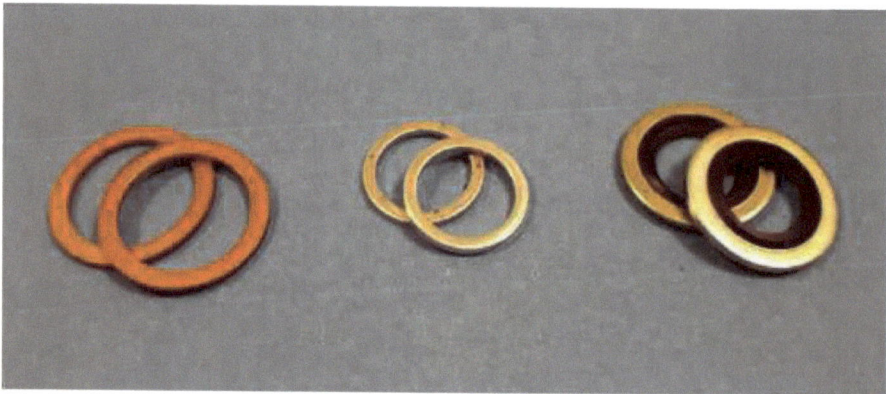

Fig. (13). Washer – Copper – Aluminum – Steel Rubber.

Circular section (o-ring) seals have also been used (Fig. **14**).

Fig. (14). Banjo connector with o-ring.

So far there have been changes in materials that have a lot to do with manufacturing costs, but there have not yet been substantial changes in designs to modify fastening operations.

Below is an example of banjo fasteners where the assembly method is improved to reduce the NVAA. The previous standard fasteners had two washers, this time one component NVAA is reduced by using one less washer (Fig. **15**).

Fig. (15). Fuel line with a copper washer.

The use of a single piece is unifies two washers in one piece, which with its bent "8" geometry can place the two necessary joints in one movement and the applied design allows the new piece to be joined to the banjo by interference fit.

Finally, here it is a new development (still in beta phase) that can be used as a perfect example of applied design, the WOW System© (Figs. **16** and **17**).

The banjo connectors of this system have the particularity that they do not need washers since the banjo is coated with a polymer that serves as a seal and reduces the NVAA in the assembly process. Another improvement by using this product is the lower number of parts needed for assembly. The manufacturing improvements are the reduction of NVAA in the assembly and the logistical improvements by reducing the quantity of different part numbers in stock.

Fig. (16). WOW System© – Banjo Fitting.

Fig. (17). Left - Washer Steel Rubber; Right - WOW System© - New Holland Tractor.

Lock Washer

Now we will look at the design applied in contemporary lock washers. Here not only the reduction of operations in assembly is considered but also functionality, after-sales and efficiency.

The vibrations and forces to which the various seals are subjected are sometimes not negligible in the automotive industry.

Over time, different types of lock washers have been applied which have not only improved the performance of the seal, but also the long-term functionality of the assembly.

A review of the different types of lock washers is:

●Washer Grower DIN 127, It is a helical washer with a cutout. It compresses when the gasket is tightened and the cutout widens to the bolt or nut on one side and to the flange on the other end (Fig. **18**).

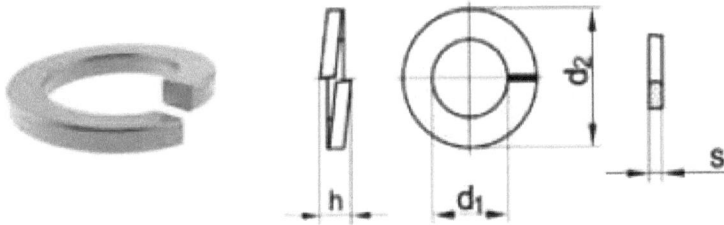

Fig. (18). Washer DIN 127.

●Washer Belleville DIN 6796; named after its creator, it is an elastic conical washer that compresses when the joint is tightened, generating a preload to the system (Fig. **19**).

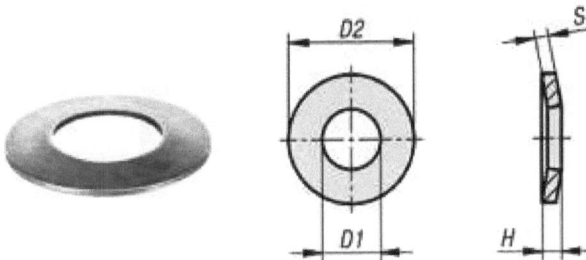

Fig. (19). Washer DIN 6796.

●Wave Washer DIN 137; is a spring washer that is also used to preload the gasket (Fig. **20**).

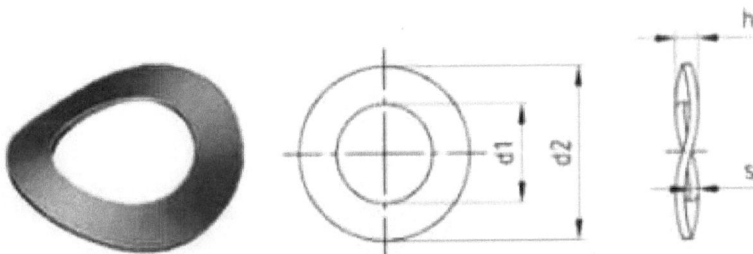

Fig. (20). Washer DIN 137.

Nowadays, improvements have been developed for this simple and important piece of engineering.

They are based on the concept of axial thrust bearings containing different plates and hydrodynamic lubrication occurs when fluid is introduced between the two plates in the form of a wedge [4] (Fig. **21**).

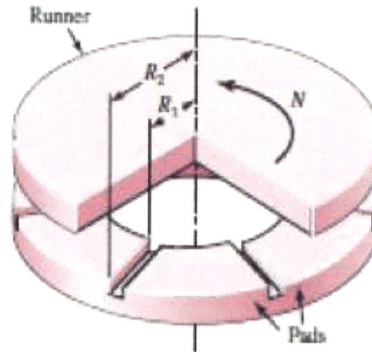

Fig. (21). Fixed-pad thrust bearing.

It is precisely this wedge format that has been designed to generate a variable geometry between the washers, where this angle differs from the helix angle of the screw thread, thus generating additional clamping force (Figs. **22** and **23**).

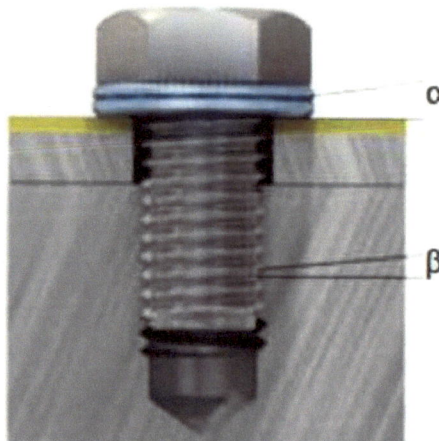

Fig. (22). The different angles - screw & washer.

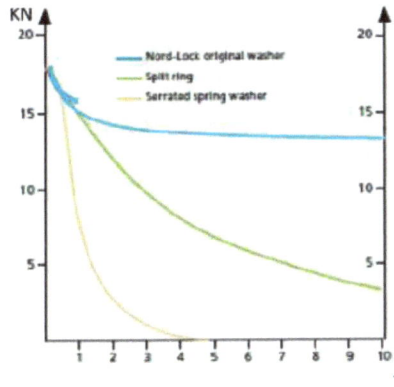

Fig. (23). The Junker test performed on steel surfaces.

An example of this are the washers developed by Nord Lock, which consist of two washers paired in a special format (Fig. **24**).

Fig. (24). Nord-Lock Washers.

Another washer developed to improve performance is presented by the Japanese company NBK (Fig. **26**), which is based on DIN 7967 lock nuts (Fig. **25**).

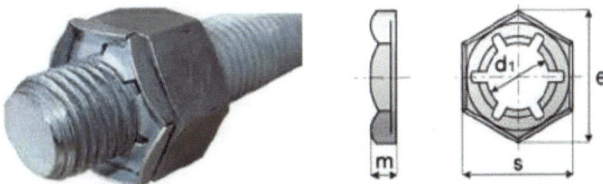

Fig. (25). Self-locking counter nuts, DIN 7967.

It takes the basis of this type of component with small modifications, which greatly improve the performance of the product.

Let us now look at these details to better understand the applied design.

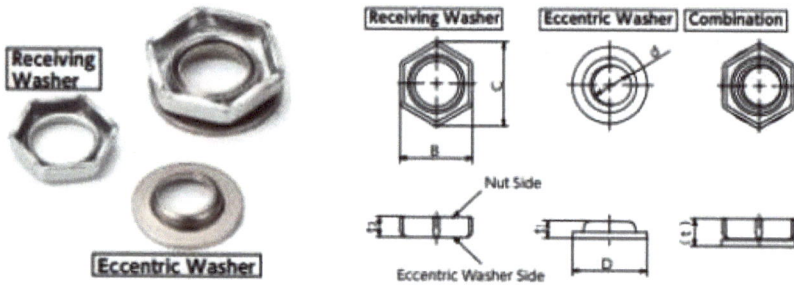

Fig. (26). SWAS EW – Eccentric Lock Washer.

These washers, unlike their predecessor, are positioned earlier and the conical part together with the eccentric hole and redirect the forces so the desired locking occurs (Figs. **27** and **28**).

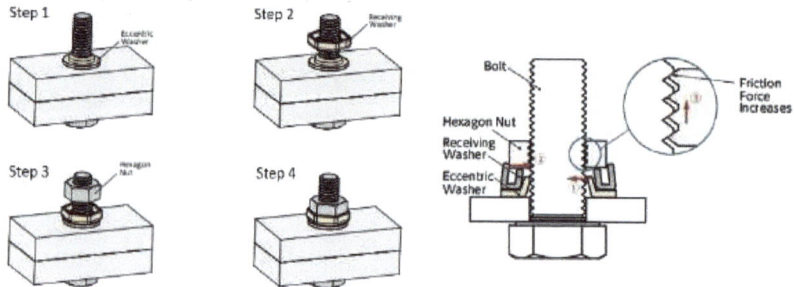

Fig. (27). SWAS EW – Assembly.

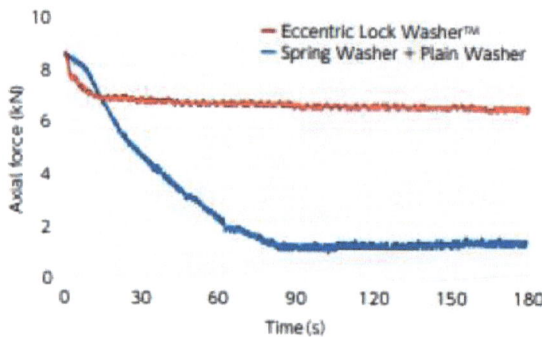

Fig. (28). The Junker test performed on steel surfaces.

CONCLUDING REMARKS

The parts and process design must be integrated, not only within a factory but also with the suppliers, thinking about the efficiency of component manufacturing, as well as assembly performance and looking at the aftermarket.

To carry out the improvements we can have 3 types of approaches: Reactive, Preventive and Proactive. For all of them we must have a feedback or a vision according to the new technologies that arise or new methods of application of these.

For Reactive approach, we must have the return information of inconveniences or problems, either from the manufacturer, from our assembly line or from our customers. It is always optimal to have early detection of problems so the impact on our customers is minimized. It is always a good practice to use standard components to reduce the downtimes.

The preventive elements will be all those in which the solutions that we give to the reactive problems are extended to obtain similar solutions before the inconvenience occurs.

Proactive topics will be those in which we will study what new technologies are in use or under development and should be incorporated to improve the performance of our product or process.

All of the above can be summarized using the expression "applied design", considering all the members of an organization, suppliers, engineering, manufacturing, customers and after-sales.

It is very important to contextualize commitment decisions as not all markets may accept new designs or processes due to different legal, regulatory, cost or market demands.

CONSENT FOR PUBLICATION

Not applicable.

CONFLICT OF INTEREST

The author declares no conflict of interest, financial or otherwise.

ACKNOWLEDGEMENTS

Declared none.

REFERENCES

[1] B.W. Niebel, A. Freivalds. Methods, Standards and Work Design 11ª Ed. DF, Mexico:Alfaomega, 2004.

[2] J. Happian, Smith . An introduction to Modern Vehicle Design. Oxford,UK: Butterworth-Heinemann, 2000.

[3] G. Boothroyd, P. Dewhurst – W. A.. Knight.Product. Design for Manufacture and Assembly – Third Edition. Boca Raton,FL: CRC Press Taylor & Francis Group. 2011.

[4] R. Budynas, J.K. Nisbett. Shigley's Mechanical Engineering Design, 8th Edition. Boston,MA: McGraw-Hill, 2008.

CHAPTER 3

Disability and the 4th Industrial Revolution

Natalia Nissen[1,*]

[1] *Department of Legal Medicine, Psychiatry and Pathology, Universidad Complutense, Madrid, Spain*

Abstract: In times where the fourth industrial revolution is looming on the near horizon, technological advances and new employment configurations invite us to question access to quality jobs for vulnerable groups such as people with disabilities.

It is observed that barriers persist in the accessibility and use of services, products and transportation and an ineffective professional training that supports the old paradigm of considering this group as a handicap in the processes of growth and industrial development. This requires investigating the management of resources, physical, technical and procedural factors involved in the design of jobs and reviewing some alternatives such as the Supported Employment methodology.

Keywords: Disability, Employment, Handicap, Industry 4.0, Individualized, Jobs, Opportunities, Skills, Supported, Work, Workplace.

INTRODUCTION

The entry into the labor market of people with disabilities entails a series of difficulties in the searching and adapting the job, mobility, maintenance of employment, and remuneration aspects.

Tackling these issues requires recognizing the differential quality of educational and vocational training programs, bonus and/or social pension systems that promote inactivity, barriers inherited from social and cultural clichés, and business barriers in hiring them.

Although different conceptions of industrialization and growth may coexist worldwide, given the level of development of each region, to speak of industries is to refer to quality of life, wealth, and development. The problem is the relationship between the accessibility chain at work and its implication in the advancement of new economic-social models and situations, opening a research

* **Corresponding author Natalia Nissen:** Department of Legal Medicine, Psychiatry and Pathology, Universidad Complutense, Madrid, Spain; Tel: +34600849234; E-mail: natinissen@gmail.com

Gonzalo F. Taboada (Ed)

field of multidisciplinary participation to provide answers in different fields of action.

As the Economic and Social Council of United Nations points out in its program "Incorporating the perspective of disability in the development program," that between 80% and 90% of disabled people of working age are unemployed in developing countries, while in industrialized countries, this figure ranges from 50% to 70%.

It shows that the greater the handicap, the greater the reluctance to employ people with disabilities, because it is believed that the tasks and duties will not be fulfilled.

In order to have a comprehensive and social perspective that takes into account the rights of people with disabilities, it is necessary to promote access to education and training, inclusive and non-discriminatory human resources policies, adequate facilities in the workplace and anti-discrimination laws [1].

Therefore, it is important to recognize that disability issues cut across disciplines and are not limited to the health and welfare professions and that participation in the industrialization process depends on factors related to rehabilitation in health as well as factors external to the organization related to people management, organizational culture and semantic issues associated with the scope of disability and its transformation [2].

CURRENT DEFINITION OF DISABILITY

The concept of disability has undergone transformations over time, giving rise to two predominant schemes that seek to define it: the individual model and the social model.

The first focuses on the individual disability and attributes any activity restriction or social disadvantage faced by the individual in his or her daily life as an unavoidable consequence of that disability.

On the other hand, the social model postulates that it is society that creates barriers for any person with a disability. These barriers include, among others, negative attitudes and inaccessible environments, systems, and structures. Disability arises when a person with a disability is excluded due to social barriers [3].

In order to unify criteria and to capture the integration of these two perspectives, the World Health Organization (WHO) established the International Classification of Impairments, Disabilities and Handicaps (ICIDH), adopted in Geneva in 1980 (updated in May 2001 as the International Classification of Functioning, Disability and Health ICIDH-2), which defines a person's functioning and disability as a dynamic interaction between health conditions (diseases, disorders, injuries, traumas, *etc.*) and contextual factors [4].

The social model has also been widely adopted and endorsed by the United Nations Convention on the Rights of Persons with Disabilities (UNCRPD) drafted in 2006 and put into force in 2008. This instrument guarantees equal participation and representation of persons with disabilities in their communities, where disability is established as an evolving concept "recognizing that disability is an evolving concept resulting from the interaction between persons with disabilities and attitudinal and environmental barriers that hinders their full and effective participation in society on an equal basis with others."

Recognizing the value of the contributions that persons with disabilities make and can make to the overall well-being and diversity of their communities, and that the promotion of the full enjoyment of human rights and fundamental freedoms by persons with disabilities and their full participation will result in a greater sense of belonging for persons with disabilities and significant progress in the economic, social and human development of society and in the eradication of poverty.

Following this definition; it is understood that the environment will become an enabler or a disabler depending on how it manages to operate in interaction with the individual.

EMPLOYMENT CHALLENGES IN THE WORKPLACE

In recent years, concepts supported by universal equal opportunity regulations have also emerged that refer to design criteria so that environments, products and services can be used by all types of users, including people with functional diversity.

An example of this are the notions of Design for All, which is the process of creating products, services and systems that are used by the greatest number of people possible, covering the greatest number of situations, and Universal Accessibility, which is understood as the condition that environments, processes, products and services, instruments, tools and devices must meet in order to be understandable, practicable and usable by all people in conditions of safety and comfort, in the most autonomous and natural way possible.

Different experiences show that if workstations are designed with general Universal Design criteria, accessibility and use would be improved in a more equitable way, simplifying any individual adaptation required by a worker with a specific limitation [5].

Equal opportunities are associated with the presence of this type of reasonable accommodation, which also includes adaptation in selection processes, evaluation of job performance and specific training in groups with functional diversity in which greater mismatches are observed in relation to the performance of tasks designed for the average population.

In turn, the variability of the difficulties makes it necessary to make an initial assessment of the job, and the worker/task/job relationship, as well as the risks derived from it on an individual basis.

The result of the analysis between labor demand and functional capability assessment is interrelated with the disruptive, variability and unpredictability characteristics of the fourth industrial revolution process [6].

This process of transformation of the sector from the automation of tasks, the high presence of cyber-physical systems that combine machinery with digital processes, the decentralization of decision making and the introduction of new forms of cooperation through the Internet of Things and cloud computing, requires a review of the components of workspaces.

New business models and startups require job skills linked to the digital world and technology, in particular professions that have to do with data analytics, statistics, and mathematics. It is here that those responsible for the design and management of work must bring into play the variables of accessibility and design for all, seeking adaptations to these models, to the productive behaviors of society and to new jobs [7].

DESIGN AND ADAPTATION OF JOBS

Making appropriate revisions for the inclusion of personnel with disabilities and responding to the demands of productivity and efficiency in the processes implies according to Mansilla Núñez (2017) [8] reviewing in which of the links of the accessibility chain redefinitions are required considering certain criteria such as:

- Integration of the work environment and accessibility of the home to the workplace.
- Necessary adaptations (physical and social) to the urban environment
- Safety and health (in the workplace and at the workstation)
- Transportation and displacement
- Communication

The complexity of these movements makes it necessary to have technicians and professionals who make use of tools to save time and resources in the evaluation and assessment of potential candidates beyond their physical, sensory or intellectual restrictions, ensuring the adaptation of the position.

To do this, when designing the position, the basic conditions are:

- Establish the set of tasks and obligations to be performed (content of the position).
- Define the working methods and procedures to perform these tasks.
- Identify a responsible person to report on their performance.

At the same time, conditions can be considered according to the perspective from which the implementation of methodologies such as Individual Supported Employment (ISE) that offer opportunities in the community and favor the use of existing ones, making the worker with a disability become more productive, independent and participative, can be evaluated.

SUPPORTED EMPLOYMENT

The methodology has its origins in the United States and Canada in the 80's, based on experimental psychology and behavior modification through reinforcement, aimed at pathologies with intellectual disabilities, currently also extends to other groups among which severe physical disabilities predominate.

One of its fathers, P. Wehman (1981) defined it as "competitive employment in integrated settings, for those individuals who have not traditionally had this opportunity, using properly trained job coaches and promoting systematic training, job development and follow-up services among others" [9].

At present, Individual Supported Employment (ISE) can be understood as employment integrated into the community within standardized companies for people with disabilities who traditionally have not had access to the labor market, through the provision of the necessary supports inside and outside the workplace, throughout their working life, and under employment conditions as similar as possible in terms of work and salary to those of another worker without a disability in a comparable position within the same company [10].

As Osburn (2006) reminds us, it is based on a structured support system and is founded on Wolf Wolfensberger's paradigm of independent living and normalization, understood as the use of the most culturally normative means possible, applied not only to the workplace but to all areas of life in which each subject can choose among acceptable options.

This characteristic allows the target group to be varied (workers with intellectual, physical, sensory, mental illness and autism spectrum disorder) [11].

The choice of support is adapted to the supports that the person needs according to the demands and requirements of the job, understanding support as the resource or strategy that allows individuals to relate to their environment (home, school, work) and increase their interdependence, independence, productivity, integration into the community and personal satisfaction.

According to Jordán de Urries (2010) [12] the fundamental points of the ISE can be summarized:

• Maintain a job in the highest number of ordinary conditions, such as salary, working conditions and safety.
• Receive the ongoing support needed to find and keep a job rather than having a person prepared for a job that may come in the future.
• Create employment opportunities rather than simply providing skills development services.
• Seek the full participation of persons with disabilities at all levels.
• Implement variable and flexible programs due to the wide range of jobs in assessing the Individual Supported Employment (ISE) methodology and the skill fit of each job.

The aim is to maintain a paid, competitive and integrated job in normalized companies with on-the-job training and the promotion of natural supports and the possibility of adapting the job (Figs. **1 – 4**)

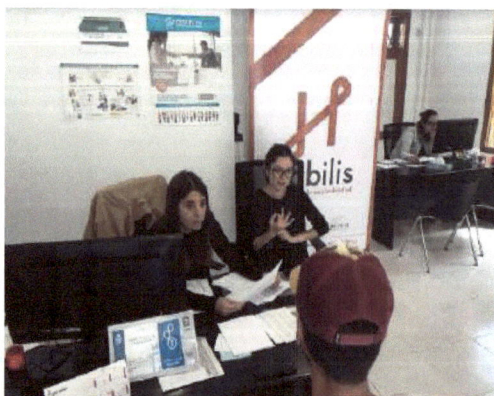

Fig. (1). Propiedad de FUNCASOR Fundacion Canaria Personas con sordera. @2020 FuncasorAqui se observa al intérprete de señas, que sería el preparador laboral, como un apoyo dentro del puesto de trabajo.

To develop it correctly, the European Union of Supported Employment (EUSE) proposes a series of phases [13].

1. Introduction to the supported employment service.
2. Drawing a professional profile.
3. Job search.
4. Company involvement.
5. Provision of supports inside and outside the workplace.

Fig. (2). Se observa el preparador laboral, *in situ*. Notas que tiene el mismo uniforme que el empleado c discapacidad. imagen propiedad de CORPORACIÓN RADIOTELEVISIÓN ESPAÑOLA, S.A RTVE.ES © Corporación de Radio y Televisión Española 2020.

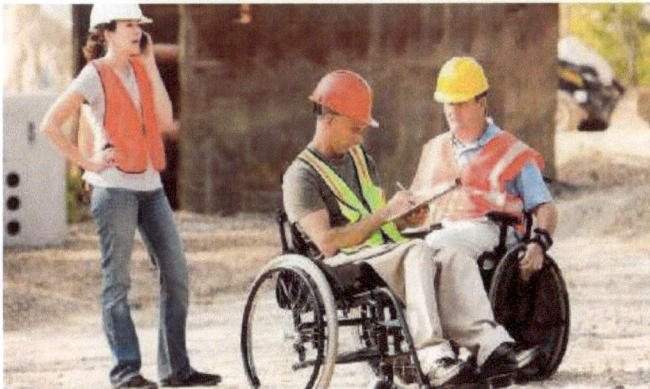

Fig. (3). Asociacion Asturiana de Transplantados cocleares y otrs problemas auditivos © 2020 Asociación ASAICPA.

Fig. (4). Aqui se observan apoyos especificos logrados a paritr de una evaluacion ergonomica del puesto de trabajo.Encontrada en https://www.interempresas.net© 2019.Interempresas Media, S.L.U.Grupo Nova Àgora.

DISABILITY AND SUPPORTS

I want to highlight the need to constantly revisit some of the interpretations of supports associated with disability. Beneath the semantic labels used are underlying concepts associated with physical abilities, cognitive functioning and/or functional deficits that influence the increased or decreased impact of disability and the acceptance and application of supports needed to perform a role in the workplace.

Thus, it is necessary to adapt categorization terms such as handicap, which sometimes excludes the presence of certain levels of abilities and has historically been limited to purely cognitive categories. The possibilities for adaptation and learning associated with personal access to work reduce the stigmatizing overtones of the concept of support.

The AAIDD (American Association on Intellectual and Developmental Disabilities) states that supports are resources and strategies that promote a person's development, education, interests and personal well-being and enhance individual functioning based on intellectual abilities, adaptive behavior, social participation, interaction, health and context.

Therefore, and returning to the ISE, the actions and use of support will be of intensive or long-term follow-up.

Intensive actions are carried out by a job coach, an external professional who accompanies the process with concrete assistance, basic and transversal competencies (training "*in situ*") and externally with complementary skills development strategies (autonomy, social relations, *etc.*).

While follow-up actions can be provided indirectly and more intermittently. Thus, the trainer gradually withdraws from the work setting as the worker's autonomy increases and the so-called "natural supports" provided mainly by co-workers come to the fore. A natural support is any person in the workplace who does not have specific knowledge about job placement but who will assist the worker with a disability in that environment.

Whether natural or specialized, support is used in a planned manner according to the intervention modality and the contract:

- In competitive employment, support is continuous, provided through a job coach, which decreases as the worker becomes proficient in assigned tasks.
- In labor enclaves the modality is mixed. The worker belongs to a specialized organization or special employment centers and is subcontracted by the company to provide temporary services.
- In groups where the contract is subcontracted through agencies to groups of workers who provide their services in standardized companies.

Whereas opting for mixed devices could contribute to the independence at work of people with Choosing mixed devices could contribute to the independence at work of people with disabilities by combining pre-employment training and on-the-job academic training and off-the-job support (outside the workplace/on-th--job), especially in medium and severe degree disabilities, allowing the activation of transversal skills and updating them to the specific need of the process.

Currently, ordinary Individual Employment with Support (IES) is evolving towards what is called Personalized Employment, with the aim of developing individualized labor relations and achieving a negotiation between the worker's possibilities and the employer's needs, seeking to reduce stereotyped jobs.

According to Griffin *et al.* (2014) [14] the advantages including customized employment are:

- Identification of specific labor rights.
- Orientation of the objectives of the individualized work.
- Response to the specific needs of the applicant and the employer.
- Identification of the person as a source of information to explore options.
- Integration of disabled and non-disabled workers into the environment.
- Facilitation of supports and funding sources.

However, among the limitations of the program, which has not yet been thoroughly researched, there is a possible tendency to generate dependence on job coaches and the negative interference it may entail in the differentiated training processes of the rest of the workers. Also, during supported employment sessions, the job coach may perform more tasks and the productivity of the person with a disability may be affected.

CURRENT CHALLENGES

The possibility of turning the work environment into an innovative training space and colleagues into natural supports at different levels of collaboration, make it necessary to contemplate these options as plastic alternatives for guidance, going from specific instructions to autonomous decision making by colleagues.

In order to guide the development of ISE programs and review compatibility with each organizational identity, it is important to establish quality standards applicable to a larger number of work environments, which would make it possible in the future to analyze the relationship between the contribution made to the different groups affected by the processes (program users, families, public and private organizations in the labor market) and the impact on companies and micro-societies in the work environment.

This reinterpretation of the methodology and further study of its effectiveness are essential for the design of lines of action for complex multipurpose processes at different scales, which do not limit the capabilities needed to perform a task to those that are easy to perform for the majority of the population.

Likewise, taking a constructivist direction in design and custom tailoring will generate new opportunities in emerging industry sectors, which are taking on new configurations as advances in technology and collaborative robotics increase and industry organizational frameworks "dialogue" with disciplines such as Machine Learning.

Finally, and in view of the aging of the population associated with a higher quality of life, the promotion of employability characteristics in an occupational model that seeks to promote active aging is a task for future research and planning.

CONCLUDING REMARKS

Regardless of the conceptualizations contemplated in each organizational culture about disability, labor inclusion is one of the central points of the 2030 agenda for sustainable development, proposed by the UN, and represents a fundamental element for full participation in the workplace.

If the life cycle of skills in the workplace requires higher rates of adaptation to change, the effective moves will be those that associate job creation in this industrial age with knowledge, problem solving based on specific experiences and personal skills (the value of which cannot be replaced by technological processes).

Thus, to base the idea of competitiveness on new elements is also to displace semantic categories and old paradigms, understanding that the only drawback of the labor inclusion process is resistance to change and not the difficulties of a group affected by the polarization of the labor market (highly qualified and well-paid jobs are created by devaluing other less specialized jobs).

It is still necessary to work so that tax incentives do not become the only incentives for access to different types of programmers and actions such as the ISE, to the detriment of productive results. The ISE is just one of the methodologies under which we seek to achieve the promotion of people with disabilities within the workplace, promoting full inclusion and the exercise of rights in a scenario where technological transformation has much to offer.

Similarly, the development of Industry 4.0 could mean a new way of organizing and achieving a change in the production processes of factories and workplaces, turning every inclusive process into an opportunity.

To this end, it is necessary to assess a broader spectrum of professional profiles, consider the time factor in the adaptation processes, flexibility in participation, access to training and condensed learning through accessible platforms.

Putting these real changes at the service of a movement of greater equality and safety in the processes as soon as the use of digital technologies within the factories and the materialization of the information acquired by the role of Big Data will accompany the development of new and specific needs that will be identified.

The near future shows that the evolution of jobs to creatively develop new challenging tasks and the interconnection of production units within factories and companies and the need for specialized training could become a single movement of constant adaptation not only for the demands of production but also for equitable and responsible social movements.

CONSENT FOR PUBLICATION

Not applicable.

CONFLICT OF INTEREST

The author declares no conflict of interest, financial or otherwise.

ACKNOWLEDGEMENTS

Declared none.

REFERENCES

[1]   D. Bas, and A. Padova, "Disability and Development Report", UN, New York, NY, 2018.

[2]   B. Choi, and A. Pak, *Multidisciplinarity, interdisciplinarity and transdisciplinarity in health research, services, education and policy: Definitions, objectives, and evidence of effectiveness.* Public Health Agency of Canada: Ottawa, Canada, 2006.

[3]   J. Campbell, and M Oliver, "Disability Politics: understanding our past, changing our future. The social model of disability: an outdated ideology? 1996", *Journal Research in Social Science and Disability,* vol. 2, pp. 9-28, 2002.

[4]   *ICIDH-2: International classification of Functioning, Disability and Health.* World Health Organization: Geneva, Switzerland, 2001.

[5]   R.L. Mace, G.J. Hardie, and J.P. Place, *Accessible Environments: Toward Universal Design.*, 1996.

[6]   J. Hernandez Galan, *Accesibilidad Universal y Diseño para Todos - Arquitectura y Urbanismo.* Artes Gráficas Palermo: Madrid, Spain, 2011.

[7]   *Study on job creation for people disabled from cloud computing.* Royal Board on Disability. Ministry of Health, Social Services and Equality Spain: Madrid, Spain, 2015.

[8]   S. Mansilla Núñez Calatrava, *La cadena de la accesibilidad y el diseño de los puestos de trabajo en la sociedad actual,* 2017.

[9]   P. Wehman, *Competitive employment: new horizons for severely disabled individuals.* Paul H. Brookes Publishing: Baltimore, MD, 1981.

[10]   K. MacDonald-Wilson, and S. Fabian, ""Best Practices in Developing Reasonable Accommodations in the Workplace: Findings Based on the Research Literature"", *The Rehabilitation Professional Journal,* vol. 16, pp. 221-232, 2008.

[11]   J. Osburn, "An overview of Social Role Valorization theory", *The SRV Journal,* vol. 1, pp. 4-13, 2006.1

[12]   M. A. Verdugo Alonso, F. De Borja Jordan De Urríes Vega, and C. Vicent Ramis, "Desarrollo de un Sistema de Evaluación Multicomponente de Programas de Empleo", *Coleccion Investigacion INICO,* 2001.

[13]   M. Evans, "Caja de Herramientas para la diversidad de la Unión Europea de Empleo con Apoyo",

[14]   C. Griffin, D. Hammis, B. Keeton, and M. Sullivan, "Making Self-Employment Work for People with Disabilities",

CHAPTER 4

People Development, Motivation & Results

Gustavo R. Mena[1,*]

[1] *Department of Psychology, Universidad Nacional de la Rioja, La Rioja, Argentina*

Abstract: The influence of strategic planning can be a link or obstacle to the good performance of the company and can be a factor of distinction in the commitment of those who are part of it.

Strategic planning allows diagnosing, analyzing, reflecting, and making collective decisions regarding current tasks and the path that organizations must explore in the future to adapt to the changes and requests imposed by the environment to achieve their viability.

Strategic planning must be understood as a participatory process, which will not solve all uncertainties, but must draw a line of those affected to act accordingly.

Keywords: Behavior, Commitment, Goals, Motivation, Planning, Roles, Satisfaction, Skills, Strategic, SWOT, Work.

INTRODUCTION

Strategic planning facilitates the possibility of thinking about the future, visualizing new opportunities and threats, focusing on the organization's mission, and effectively guiding its course, facilitating direction and leadership. It also allows facing problems such as the allocation of human and financial resources. In relation to classic forms of management, strategic planning introduces a modern methodology.

This modern form of management requires deep knowledge of the organization, greater participation, improving communication and coordination between different levels, and improving management skills, among others.

For the reasons mentioned above, the investigation raises the following question.

How does strategic planning influence employees' job performance?

[*] **Corresponding author Gustavo R. Mena:** Department of Psychology, Universidad Nacional de La Rioja., La Rioja, Argentina; Cell.: +54 9 351 5184993; E-mail: gustavo_mena17@hotmail.com

Gonzalo F. Taboada (Ed)

Nowadays, due to the changes and the new technologies that appear in the market that determine the development of more abilities, skills, and knowledge, organizations find themselves with the need to implement changes in their labor strategy when facing the challenges presented to them.

It is necessary for companies to develop new techniques for production, market, distribution, service, and customer service, for which human capital is needed, and thus assume the organizational challenges.

Within this context, the productivity and management of human capital in organizations will become key elements of survival; therefore, coordination, direction, motivation, and staff satisfaction are increasingly important aspects of the administrative process [1,2].

PLANNING

Planning is the process by which the company's executive bodies continually design the desirable future and select how to make it feasible; It is based on advanced decision-making that, in a systematic and complex way, is oriented to ensure the highest probability of achieving previously designed desired futures.

The planning process works as a global system, considering that all the functions and organizational levels of the company must be planned simultaneously and interdependently, using the methodology of systemic reasoning since the global system works through the joint interaction of its components. Planning is usually classified as strategic and tactical.

Technical planning aims to optimize the allocation of resources to the maximum consequence of compatible company objectives. It is known as programming and is linked to the third stage of budgeting, which refers to the specific time frame in which the accepted programs must be executed. Within the integral perspective of the planning process, budgets are the formalized quantitative and qualitative expression of the partial or global quota of the programs that must be executed in each period.

Three classes are distinguished:

a. Optimal, which follows policies of maximization or minimization of the variables.
b. Satisfactory, which refers to the level of objectives that, by consensus, satisfy the overall organization of the company.
c. Adaptive, which emphasizes the planning process itself, highlighting its value for training and continuous learning.

The plan represents the set of explicit and coherent decisions to allocate resources to pre-established ends.

Planning represents the exercise (the concrete application) of the plan linked to the theoretical instrumentation required to transform the economy or society; for this reason, it is the concept of planning that will be taken during the investigation.

Strategic planning is nothing more than the process of relating the goals of an organization, determining the policies and programs necessary to achieve specific objectives on the way to those goals, and establishing the necessary methods to ensure that the policies and programs are executed, that is, is a formulated long-term planning process that is used to define and achieve organizational goals [3].

Its characteristics are:

a. It is conducted or executed by high hierarchical levels.
b. Establish a framework for the entire organization.
c. It faces higher levels of uncertainty with respect to other types of planning.
d. Generally, it covers long periods. The longer the period, the more irreversible the effect of a more strategic plan will be.
e. Its parameter is efficiency.

The central objective of strategic planning is to get the most out of internal resources by selecting the environment where they are to be deployed and their deployment strategy. For example, it is about finding a market niche that the company can serve better than potential competitors; therefore, the application of resources is more beneficial than in other circumstances.

It is the antithesis of improvisation and a management style that reacts instead of making decisions based on a plan. Due to its characteristics, it allows the manager to grasp the future of his organization as it commits him to follow previously defined lines of action.

The intensity of planning will depend on different factors such as the anticipation with which a decision must be made, the difficulty of coordination in decision-making, the magnitude of deviations and the consequent changes to be introduced. Available resources cannot be ruled out.

The limits are defined according to the type of organization, the behavior of resources, and the type of activity in question. The short, medium, and long terms can be distinguished.

The long term encompasses items subject to planning such as products, profits, return on investment, cash flow, development, and staff training plans.

The medium term is the period in which the long-term plans are expressed in more detail: the plans for profits, sales, production, inventories, expenses, purchases and financial statements.

In the short term, in terms of planning, maximum detail is reached in everything related to planning: level of installed capacity, fixed costs, variable costs and others.

Problems due to lack of planning can be:

Ignorance of the company's progress, lack of control, only works in the short term, today there are no guidelines for action. There are no clear criteria for decision making and it is not possible to project the future of the organization, and critical situations get out of control.

To define any planning, organizations must first be clear about their mission and vision.

The Mission will be the raison d'être of the company and the Vision is the projection that gives orientation and strategic sense to the decisions, plans, projects, *etc.*

Another very important point during strategic planning is to carry out an analysis that allows to know those internal and external factors that can help or hinder the organization's operation.

A recommended tool for detecting and classifying the types of opportunities and threats is the "SWOT" analysis (Fig. **1**).

Strengths: These are the characteristics of the company that favor the achievement of its objective.

Weaknesses: They are those characteristics that constitute an internal obstacle to achieving the objectives.

Opportunities: These are those situations in the environment that would favor the achievement of the objectives.

Threats: Situations that arise in the company's environment that could negatively affect the possibilities of achieving the objectives.

Strategy selection can have several approaches, according to a traditional approach is the determination of the basic long-term goals and objectives in a company, together with the adoption of action plans and the allocation of resources necessary to achieve these purposes, always considering the boundary conditions or context. The basis of this approach is that a strategy is a means and an objective will be an outcome.

The implementation of the strategy should consider 4 components.

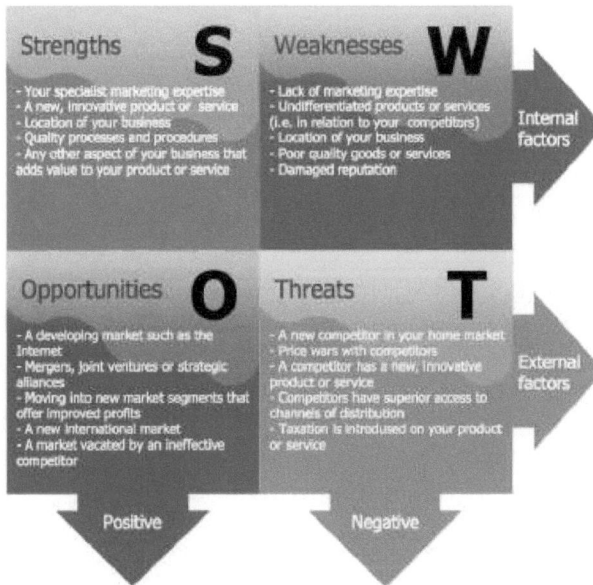

Fig. (1). SWOT Analysis - created in the 1960s by Albert Humphrey at the Stanford Research Institute.

- Designing an organizational structure: the organization, in order to achieve the functioning of a strategy, needs to adopt the right structure, this implies assigning responsibilities for tasks and authority for decision making, analyzing the best way to do it (dividing the organization into subunits, distributing authority, distributing authority among different hierarchical levels, achieving integration among subunits).
- Design of control systems: the organization must decide on the best way to evaluate performance and control the actions of the sub-units, from market and production controls to bureaucratic and control alternatives through organizational culture, it also implies deciding on the type of compensation and incentive systems to establish for its employees.
- Matching strategy, structure and controls: Because different strategies and environments place different demands on an organization, they require different

responses and structural control systems. If the company wants to be successful, it must find a balance between its strategy, structure and controls.

- Managing conflict, politics, and change: Organizational politics plays a key role, is endemic to organizations and different subgroups have their own agendas and conflicts, such as competing for a greater share of the organization's finite resources. organization. These conflicts can be resolved through the relative distribution of power among subunits or through a rational assessment of relative need.

Power struggles and coalition building are the major consequences of these conflicts and are, in fact, part of strategic management.

Strategic change tends to highlight these struggles, since any change causes an alteration in the distribution of power within an organization.

Once the strategy has been implemented, its execution must be monitored to determine the extent to which the strategic objectives are being achieved. This information is fed back to the corporate level through feedback loops, which provide the next phase of strategy implementation and formulation, serving to reaffirm existing corporate objectives and strategies or to suggest changes.

The combination of strengths and opportunities gives rise to potentials, which indicate the most promising lines of action for the organization and make it possible to detect and generate a competitive advantage.

A company is said to have a competitive advantage when its profit ratio is higher than the average for its industry.

The profit rate is usually defined as a given index and the fundamental determinant of a company's profit is its gross profit margin (GPM), which corresponds to the difference between total revenues (TI) and total costs (TC), divided by total costs: $GPM = (TI-TC) / TC$.

For a gross profit margin to be higher than the industry average, one of the following must occur:

- The company's unit price must be higher than that of the average company and its unit cost must be equivalent to that of the average company.
- The company's unit cost must be lower than that of the average company and its unit price must be equivalent to that of the average company.
- The company must have a lower unit cost and a higher unit price than the average company.

This does not contemplate monopolistic practices that distort the market.

Therefore, for the company to gain a competitive advantage, it must either have lower costs than its competitors or differentiate its product so that it can charge a higher price than its rivals or it must carry out both options simultaneously.

When a company charges a higher unit price than the industry average, it enters a higher price range, but for the consumer to purchase it, the company must add value to the product, from the consumer's point of view, and this requires differentiating the product from those offered by rivals in one or more dimensions (quality, design, delivery time, .).

The most common factors that constitute competitive advantages are: Efficiency, Quality, Innovation and Customer Service. These are the building blocks of competitive advantage. These bricks are generic in the sense that they represent four basic ways to reduce costs and achieve differentiation that any company can adopt, regardless of its industry or the products or services it offers.

1. - Efficiency: Outputs are the goods and services generated by a company. Efficiency is measured by the cost of inputs required to generate a given output. The more efficient an organization is, the lower the cost of inputs needed, so efficiency helps a company achieve a low-cost competitive advantage. One of the keys to achieving high efficiency is to use inputs as productively as possible. The most important component of efficiency for most companies is worker productivity, which is usually measured by considering output per employee. Holding this condition constant, the firm with the highest productivity per worker in an industry will typically have the lowest production costs. In other words, that organization will have a competitive advantage based on its costs.

2. - Quality: Quality products are reliable goods and services in the sense that they fulfill the function for which they were designed, perform well and at the lowest cost. The first impact of high product quality on competitive advantage is twofold. Offering high quality products creates a brand reputation for a company's products. In turn, this enhanced reputation allows the company to charge a higher price for its products. The second impact comes from higher efficiency and thus lower unit costs caused by higher product quality. In this case, the biggest effect is the impact of quality on productivity. Higher product quality means that less time per worker is spent manufacturing defective products or providing unusual services, and less time is spent correcting errors. This translates into higher productivity per worker and lower cost per unit.

3. - Innovation: Innovation can be defined as something new or novel with respect to the way a company operates or the products it generates. Therefore,

innovation includes advances in the types of products, production processes, administrative systems, organizational structures and strategies developed by an organization. Although not all new products are successful, those that are can be an important source of competitive advantage. The reason is that successful creation provides a company with something unique, something that its competitors do not have. This exclusivity can enable a company to differentiate itself from its rivals and charge a higher price for its product.

4. - Customer Service: To achieve customer acceptance, a company must provide customers with exactly what they want when they need it, *i.e.*, it must do everything possible to identify and satisfy their needs. Among other things, achieving a greater ability to correspond with the customer involves providing them with the value of what they have paid for. In addition, satisfying customer needs may require the development of new products with features that existing products do not possess. In other words, achieving greater efficiency, quality and innovation are all part of achieving great customer service. Another factor that stands out in any analysis of customer responsiveness is the need to customize goods and services to individual customer demands. One aspect of customer responsiveness that has generated increased attention is customer response time, the time it takes to deliver a good or provide a service.

All of these factors increase customer compliance and allow a company to differentiate itself from its competitors and build brand loyalty and price its products higher.

The benefits of strategic planning enable the organization to define its future, *i.e.*, the organization can undertake and influence activities (rather than merely respond, *i.e.*, it has a preventive/proactive position) and, therefore, can control its destiny.

The greatest benefit of strategic planning has been that it helps organizations to have better strategies, because they use a more systematic, logical and rational approach to choosing their strategies. There is no doubt that this is an important benefit of strategic planning, but there has been some research that shows that the most important contribution of strategic planning is in the process, not in the decision or the resulting document.

The way in which strategic planning is carried out takes on great importance. One of the central objectives of the process is that all managers understand and commit to it.

Understanding may be the most important benefit of strategic planning, followed by commitment.

When managers and employees understand what the organization does and why, they tend to feel part of the company and are committed to helping it. This is especially true when employees also understand the links between their personal compensation and the organization's performance.

Thus, one of the great benefits of strategic planning is that the process offers the opportunity for empowerment, *i.e.*, giving staff the power to decide. The act of empowerment reinforces the sense of personal efficacy.

Financial Benefit: Companies that achieve optimal results tend to plan simultaneously to prepare for future fluctuations in their internal and external environment. Companies with planning systems that most closely resemble strategic planning theory tend to achieve superior long-term financial results compared to their industry or sector.

Companies with superior performance appear to make more informed decisions and anticipate short- and long-term consequences very well. In contrast, poor performers tend to engage in activities that do not reflect well on forecasts of future conditions.

Non-Financial Benefits: In addition to helping companies avoid financial problems, strategic planning offers other tangible benefits; for example, greater alertness to external threats, better understanding of competitors' strategies, increased employee productivity, less opposition to change, and a clearer understanding of the relationship between compensation and performance.

Strategic planning strengthens the ability of organizations to prevent problems by fostering the interaction of managers and functions at all levels.

Strategic planning, in addition to empowering managers and employees, often imposes an order and discipline on the company that can be the beginning of an effective and efficient management system. Strategic planning can either renew confidence in the current business strategy or signal the need for corrective action.

The strategic planning process lays the foundation for all managers and employees in the company to identify and rationalize the need for change, *i.e.*, it helps them to see change as an opportunity rather than a threat.

It offers a cooperative, integrated, and enthusiastic approach to attack problems and opportunities.

Human Resources Planning

It is the process of anticipating and preventing the movement of people in and out of the organization. Its objective is to use these resources as effectively as possible, where and when they are needed, to achieve the organization's objectives.

Human Resource Planning (HRP), also called workforce planning, is a process that allows us to "put the right number of qualified people in the right position at the right time."

On the other hand, we can also define Human Resource Planning as the system that allows adjusting the supply of internal personnel (available employees) and external personnel (employees who are looking for or will be hired) to the vacancies that the organization expects to have in a given period.

The basic objectives of human resource planning are as follows:

- Optimizing the company's human factor.
- Ensure timely, qualitative, and quantitative staffing of the necessary personnel.
- Develop, train, and promote current staff, according to the future needs of the company.
- Motivate the human factor of the company.
- Improve the work environment.
- Contribute to maximize the company's profit.
- Generate commitment and a sense of belonging to the members.

A very important element that must be analyzed when carrying out a strategic planning process within the subsystems that make up the organization is the size of the structure.

The structure dimension considers the interrelated system of officially sanctioned roles that are part of the organization chart, the definition of roles and responsibilities. Roles are defined, interrelated and responsibilities are designated accordingly.

The structure is the consequence of the requirements of the strategy to achieve the objectives, it is efficient in terms of activities, functions, roles, delegation of authority, control section and other components that achieve the expected results based on the objectives. The structure is a consequence of planning to establish and maintain the appropriate relationships between tasks and their groups to totalize the effort of the staff to achieve the planned objectives, allowing the following:

- Define, discuss, and evaluate objectives.
- Set responsibilities.
- To have a total vision of the organization.
- Train staff.
- Distribute tasks.
- Leverage staff capacity.
- Budget staffing needs over time.
- Identify personnel with the organization.
- Eliminate duplication of efforts.
- Shorten lines of communication.
- Eliminate unnecessary functions.
- Enable the organization to adapt to change.
- Make better decisions.

The structure responds to factors such as: size of the organization, complexity, formalization, .

The size or dimension of the organization is the number of members of an organization or the scale of operation.

In terms of its importance, size is the key to understanding what happens to the organization and its members.

Large size is related to an increase in complexity, in terms of specialization and horizontal and vertical differentiation, is directly related to an increase in formalization, can be deduced from the size in which other factors are involved: technology, environment and personnel characteristics.

Large organizations tend to have more specialization, departmentalization, more vertical levels, more rules and procedures. Therefore, they have greater difficulty in communication, control and coordination, but have more power over the environment, resources for planning and less dependence on individuals.

Complexity is determined by division, separation, and specializations.

The more complex, the greater the coordination, control, and communication effort.

The three elements of complexity are:

- Horizontal differentiation: is the way in which the task is divided among the members determined by the nature of the task, the degree of structuring of the task and the routine. There are two basic forms: Specialist in specialized tasks

and Specialist in partial tasks.
- Vertical differentiation: the number of supervisory levels or the average number of hierarchical levels.
- Spatial arrangement: this can be a form of vertical or horizontal differentiation, the activities of the personnel are physically dispersed.

Formalization is the prescription of how, when and by whom a task should be performed. It can be rigid or flexible, written, or unwritten. It is the degree of standardization of work.

The important thing about formalization is the observance of rules and procedures designed to handle organizational contingency.

If formalization is at a maximum, it means that contingencies were foreseen. At the other extreme, when it is minimal, in the case of specific situations, if they are repeated, the procedures are formalized, which implies organizational learning.

The formalization of the procedures, the corporate actions, and the determination of roles are transmitted through a manual of functions, being this a work instrument that contains the set of rules and tasks that each employee develops in their daily activities and that will be technically elaborated based on the respective procedures, systems, norms and that summarizes the establishment of guidelines and orientations to develop the routines or daily tasks, without interfering in the intellectual capacities, These will be technically elaborated based on the respective procedures, systems, norms and that summarize the establishment of guidelines and orientations to develop the daily routines or tasks, without interfering in the intellectual capacities, since they will be able to make the most correct decisions supported by the superiors, and establishing clearly the responsibility, the obligations that each one of the positions entails, their requirements, profiles, including the work reports that must be elaborated at least annually within which the qualitative and quantitative work done in the period is indicated, the problems and inconveniences and their respective solutions both in Forms and manuals must be permanently evaluated by the respective bosses to guarantee their correct development.

What determines the number of formalizations is the nature of the task, the technology, the conditions of the external environment, the nature of the personnel, .

A high degree of formalization leads to bureaucratization of the employee or to fight it. The subject may feel imprisoned by formalities if the formalization is excessive or disorderly if it is ambiguous.

Within an organization, different types of structures can be found according to Brown and Jaques. They point out four coexisting organizational forms and the same institutional environment.

- Formal or manifest structure: it is the situation as it is officially described or shown, whatever the apparent situation may be.
- Presumptive structure: what members assume, perceive, or fantasize.
- Existing structure: it is the one that operates and can be inferred through the systematic analysis of the actual variables and processes, *i.e.* the real one.
- Required structure: it is the situation as it should be to satisfy the needs, the desirable one.

The final stage of planning is strategic control management, which involves interpreting and evaluating the behavior of external variables and how the organization has responded to measure deviations. Corrective decisions must be made before a process is completed. It is carried out under conditions of uncertainty since measures must be taken according to the evolution of the environment and how it influences progress towards the achievement of objectives. Strategic control is configured as the process of regulating and controlling the system to ensure the success of the strategies formulated.

The objective of strategic control is to know and monitor the evolution of the environment, competitive forces, and the effectiveness of the organization in the implementation and achievement of the objectives contained in the strategies designed.

There are four specific aspects: monitoring the future, observing the progress of strategic action, monitoring key external variables and reviewing all stages.

The fundamental objective at this level of control is to achieve:

- Evaluate the effectiveness of the strategic plan and check the extent to which the objectives are achieved.
- To globally identify the results obtained.
- Evaluate pure strategic, managerial, and operational management with a global approach.
- Design an information system referred to in the literature as "early warning and response" that is accessed in a timely manner and in real time.
- Verify whether the organizational structure, systems and procedures are adequate for the implementation of the strategy.

In defining the objectives of strategic control, the research process will focus on it as the final stage of planning, considering both the individual activity of each department and the influence of control on the overall objectives of the organization.

This set of control indicators makes it possible to orient and subsequently evaluate the contribution of each area to the organization's objectives and to make a predictive model that makes it possible to estimate the result of the activity to be carried out and, in case of deviations, to take corrective decisions.

In order to take advantage of the opportunities offered by the environment, enhance their capabilities and resources, the result of the control must respond to an appropriate organizational structure that helps the performance of the employee's work and allows an effective communication system.

Communication acquires its full value when it responds to an overall program that is fully integrated into the strategic plan of each company or institution. What, when, where and how to communicate are the questions to be asked before starting to professionally use the vast resources of communication.

Strategic communication should be understood as a participatory process that allows us to draw a line of purpose that determines how the objectives will be achieved.

Strategic communication requires adequate planning, understood as the process by which an organization, once it has analyzed the environment in which it operates and set its short- and long-term objectives, selects the most appropriate strategies to achieve these objectives and defines the projects to be executed for the development of these strategies.

When aiming for effective strategic communication, the strengths and weaknesses of the organization's internal environment must also be taken into account, especially determining what the organization can do with the means and resources available, as well as the elements of the internal structure that might seem inadequate or insufficient when dealing with increased demand from external audiences.

If there is insufficient alignment between the organization's mission, its capabilities, and the demands of the environment, it will be an organization that does not know its real usefulness. Thus, an effective strategic plan helps to balance these three forces, to recognize potentialities and limitations, to take advantage of challenges and to face risks.

Fundamentally, a strategic communication plan produces benefits related to the ability to carry out a more efficient management, freeing up human and material resources, as well as promoting the participation of the recipient, so that he/she is directly involved in the process.

A strategic communication plan is a proposal of communication actions based on concrete data, objectives, and budgets. This plan is a branch of the organization's marketing plan, so they must go hand in hand and can never contradict each other, on the contrary, they must obey the institutional policies and its mission and vision.

There are many achievements that can be obtained with good communication and that this is only possible by applying it through a properly structured strategic plan. Among the achievements the author mentions: coordination, motivation, and management facilitators.

Coordination allows the synergy of the different parties involved in a project, since with cooperative integration, strategic objectives are achieved more effectively.

In terms of motivation, it is stated that knowing the project and seeing the place that each member will occupy encourages them to project their desires, facilitating adaptation to the changing environment in which the organization is immersed [4,5].

Motivation

The performance of the members of an organization is directly related to its functioning and its success and future will depend on it.

Some authors consider that individual characteristics, such as capabilities, skills, needs and qualities, interact with the nature of the work and the organization to produce behaviors that can affect the results and unprecedented changes that occur in organizations.

The importance of this concept lies in the fact that a worker's behavior is not the result of existing organizational factors but depends on his or her perceptions of these factors. However, to a large extent these perceptions may depend on the interactions and activities, as well as other experiences of each member with the organization.

Between recognition within the organization and the satisfaction of their needs, their motivation will become the driving force to assume responsibilities and orient their work behavior towards the achievement of objectives that will allow the organization to reach high levels of efficiency and work performance.

Campbell, Dunnette, Lawler and Weick (1970) distinguished between behavior, performance, and efficacy. According to these authors, behavior is what people do, and does not include any evaluative component. Performance refers to the members' contribution to the organization's objectives. In this case, one can speak of better or worse performance to the extent that the subject's behavior does or does not contribute to the smooth functioning of the organization. Campbell insists on two characteristics of performance [6].

1. Performance includes only those behaviors or actions that are relevant to the organization's objectives. A controversial issue is who decides which behaviors are relevant or contribute to achieving the organization's objectives.
2. Performance does not denote the consequences or results of the action, but the action itself. There are certain behaviors that affect the achievement of the organization's objectives and yet are not observable (*e.g.*, thinking). In this case, performance can only be deduced from the results of the behavior.

When the contributions of these members are translated into results, we speak of "effectiveness" or "efficiency" depending on the characteristics and performance of the task. It is therefore important to define and differentiate between the two terms.

Efficacy is the ability to achieve the expected or desired effect after performing an action. Effectiveness depends not only on the contributions of the subjects, but also on several factors beyond their control. A person can show good performance, to the extent that he or she directs his or her actions with effort and actively towards achieving the organization's objectives, but still not be very effective.

Efficiency refers to the rational use of means to achieve a predetermined objective (*i.e.*, to accomplish an objective with the minimum of available resources and time).

Beyond the difference, the combination of effectiveness and efficiency is the ideal way to achieve a goal or objective. Not only will the desired effect be achieved, but the least amount of resources will have been invested to achieve the results.

During the task or once it is completed, it is necessary to evaluate it, to determine what degree or level of performance the employee will achieve.

Performance appraisal is a tool for improving the performance of the company's human resources, regardless of whether there is a formal appraisal program in the organization. Line managers always observe how employees perform their tasks and form impressions about their relative value to the organization.

When a person joins an organization, after a long recruitment and selection process, the company must be concerned with creating the conditions that allow the new employee to generate a commitment that reflects intellectual and emotional involvement with the company, that allows him/her to function at his/her maximum potential, to achieve greater productivity and to help achieve the organizational objectives. As in all human groups, something is always expected from the other, that is, there is an expectation about what the other is going to do, how he/she is going to do it and why he/she is going to do it.

The concept of work commitment is identified in organizations as the bond of loyalty or belonging by which the worker wishes to remain in them, due to their implicit motivation. Commitment as a process of identification and belief in the importance of their work, the need and usefulness of the functions they perform at work.

Job commitment: can be defined as the degree to which a person identifies with his or her job, actively participates in it and considers his or her performance to be important for his or her own evaluation. From which we derive the concept of job satisfaction, which is the general attitude of an individual towards his or her job.

Organizational commitment: the degree of identification with an organization, its objectives, and the desire to remain in it as one of its members.

That said, there is a human-organization relationship known as the Psychological Contract.

According to Edgar Schein, the psychological contract emphasizes that alongside the explicit (usually written) contract by which every individual who enters an organization establishes a formal bond with the organization, which are the normative specifications of what the organization expects of him and what the organization provides him, an implicit ("Psychological") contract is established in which both parties place expectations, desires and aspirations in each other that they expect to be reciprocally fulfilled.

If expectations are not met or the exchange is not perceived as fair, dissatisfaction and conflict arise. If it is the organization that is disappointed, the decision may be dismissal, marginalization, deferral, or sanction for "non-compliance". If it is the individual, the decision may be resignation, rebellion, apathy and others.

If the organization fulfills only the formal contract, but not the psychological one, workers tend to perform less well and have less job satisfaction, since their intrinsic expectations are not met. On the contrary, if these are fulfilled both economically and psychologically, workers feel satisfied, stay in the organization, and have a high level of performance.

As needs and external forces change, expectations also change, making the psychological contract a dynamic contract that must be constantly negotiated. The psychological contract is a powerful determinant of organizational behavior, even though it is not written down anywhere.

Motivation is an aspect of crucial importance in determining an individual's performance since the degree of motivation depends on the individual's performance and effectiveness at work.

Accordingly, there is a certain level when considering behavior depending on the person's ability to perform it and his or her motivation to do so. Therefore, one of the factors that determine a person's job performance is motivation.

Motivation represents the action of active or driving forces, it is only partially understood, it involves needs, desires, tensions, discomforts, and expectations. It implies that there is some imbalance or dissatisfaction within the relationship between the individual and his environment: he identifies objectives and feels the need to carry out certain behaviors that vary from one individual to another, both values and cognitive systems, as well as skills to achieve personal goals, these needs, values and personal capabilities vary in the same individual over time, it is subject to the stages that man goes through, from childhood he seeks to complete his studies or when the worker has other aspirations and motivations.

The process that stimulates human behavior is similar in all people, despite the differences mentioned above. But there are three premises that explain human behavior.

- Behavior is caused: that is, there is an internal or external cause that provokes human behavior, due to heredity and environment. Behavior is caused by internal and external stimuli.
- Behavior is motivated: In all behavior there is an "impulse", a "desire", a "need", a "tendency", exposures that serve to indicate the reasons for the behavior.
- Behavior is goal-oriented: In all human behavior there is a purpose, since there is a cause that generates it. Behavior is neither causal nor random, it is always directed and oriented towards some goal (Fig. **2**).

Fig. (2). Human Behavior.

The objectives, the satisfaction of needs and a positive mentality towards the work activity, become the fuel that activates motivation, and this provides the necessary disposition for a better development of the work activities.

The managers of an organization must understand that the behavior of their employees is driven by their motivation, therefore, it is necessary to understand that it is this motivation that mobilizes the members to make the maximum effort to achieve the proposed objectives and at the same time satisfy some individual needs.

Motivation may be due to intrinsic factors, which are related to job satisfaction and the nature of the tasks the individual performs.

Therefore, motivational factors are under the control of the individual, as they are related to what he/she does and performs. Motivational factors involve feelings related to:

• Personal growth and development.
• Personal recognition.
• Self-actualization needs.
• Greater responsibility.

On the other hand, motivation can be caused by factors extrinsic to the worker, that set of conditions related to the work context. It is in the environment that surrounds people and encompasses the conditions in which they perform their work.

This type of motivation is activated by needs, values, goals and cognitions developed from motivational aspects that are not specific to the work activity, which the person carries out in order to achieve them and for which he/she experiences external control (Fig. **3**).

Fig. (3). Motivational Cycle.

In this case the extrinsic motivation includes:

- Salary.
- Job security.
- Working conditions.
- Social benefits and services.
- The prestige.
- The quality of relationships.
- Interpersonal.
- The quality of supervision.
- The company's procedures.

It is observed that the motivation process has a beginning, but no end, since it is a cycle that feeds back, as new needs arise. The person is in a permanent state of motivation, and as one desire is satisfied, another arises in its place.

In short, motivation is a variable of job performance, the more motivated a person is towards something, the more effort he will put into achieving it; the more reasons he finds to do a job better, the more effort he will put into it.

In addition to the satisfaction of basic needs, goals, the desire for achievement and improvement, as well as the need for self-fulfillment can be powerful reasons for seeking optimal performance. On the way to achieving their goals, individuals grow, goals become tools for people's development; only someone who has no goals will get nowhere.

Motivation is an aspect of crucial importance in determining an individual's performance, since his or her performance and effectiveness at work will depend on his or her motivation.

All work is a manual, mental or mixed activity aimed at satisfying needs, transforming the environment, and maintaining adequate mental health. Therefore, we can deduce that: if work is a means to satisfy needs, these constitute reasons to work.

Consequently, there is a certain level of agreement in considering behavior as a function of people's ability to perform and their motivation to do so. Therefore, one of the factors that determine a person's job performance is motivation.

Work performance variables:

- Working conditions
- Degree of training
- Experience and technification
- Physical and emotional health
- Degree of collaboration between colleagues, managers, supervisors, .
- Degree of motivation towards the activity and/or the fruits it produces.

The functional relationship can be described as directly proportional and can be taken as constant, the goals the individual sets for himself, as well as his enjoyment of the work he performs.

CONCLUDING REMARKS

When talking about strategic planning, it is necessary to first identify the components that make up the project, among which we can mention the objectives, mission, and vision.

What is important is the clear and massive dissemination of the objectives, since different members of the organizations may have different interpretations, and this may lead to inefficiencies when allocating resources to them.

The company's vision and mission must also be conveyed by all possible means.

Another important issue is ongoing training, which is the best investment that companies have, as it is a fundamental part of motivation, beyond strictly salary-related issues.

It is always very important, for the motivation of the team, that all the information of the company is expressed in a positive way and maintaining an expectation that allows to think about growth.

CONSENT FOR PUBLICATION

Not applicable.

CONFLICT OF INTEREST

The author declares no conflict of interest, financial or otherwise.

ACKNOWLEDGEMENTS

Declared none.

REFERENCES

[1]   J. Rodriguez Valencia, "Cómo aplicar la planeación estratégica a la pequeña y mediana empresa",

[2]   I. Chiavenato, "Administración de Recursos Humanos, 5ta",

[3]   Mintzberg, Quinn, and Voyer, "EL Proceso Administrativo – Conceptos, Contextos y Casos- Edición Breve",

[4]   S. Cane, "Cómo triunfar a través de las personas",

[5]   J.M. Peretti, "Todos somos directores de Recursos Humanos", *España: Gestión 2000. ,* 1997.

[6]   H. Mintzberg, "Mintzberg y la Dirección", In: *España: Editorial: Díaz de Santos S. A. ,* 1991.

CHAPTER 5

Low Cost Logistics Solutions

Cristian F. Stiefkens[1,*]

[1] *Industrial Compass, Córdoba, Argentina*

Abstract: In this article, we are going to review different solutions for material handling problems to increase productivity.

In this process, we learn about some tools used to analyze time and methods and improve them.

The aim is to demonstrate that automation is not always expensive; if we resort to our knowledge, we can find a mechanism cheaper than a robot.

Keywords: Automation, Chart, Design, Diagram, Karakuri, Low-cost, Manufacturing, Movement, Payback, Work, Yamazumi.

INTRODUCTION

During my professional life, I have seen many automation projects that have not been fully developed because they did not have a successful payback period. I have also seen several ideas that were discouraged before they were implemented as well as projects that did not deliver the expected results.

After studying the principles of World Class Manufacturing, I realized that the problem is not automation. When we hear this word, the first thing that comes into our mind is "robot". In fact, that is only a limited view of the concept because we can automate processes with a lot of low-cost automation mechanisms. The key element is to find out which of these automation processes is necessary and consequently, apply it accurately.

For example, in a project to automate the end of a packing line there were 6 lines where different products were packed through the same process. A filling machine inserted the products into the primary container, then two operators grouped the

---

* **Corresponding author Cristian F. Stiefkens:** Industrial Compass, Córdoba, Argentina; Tel: +5491160009799; E-mail: cstiefkens@hotmail.com

Gonzalo F. Taboada (Ed)

filled units and put them into a box. After that, another operator took the boxes and consolidated a pallet. Once the pallet was completed, it was wrapped and stored (Fig. **1**).

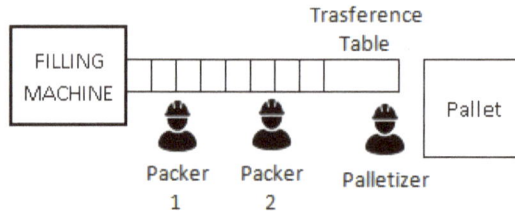

Fig. (1). Standard Work Without Automation.

The automation proposal was to buy 3 machines, one to prepare the box to receive the product, a robot to pick the product and place it into the box, and another robot to palletize the box. The investment was huge, but the process was not fully automated because a skilled operator was required to feed and operate the 3 machines. In addition, productivity had to be improved so we had to look for another solution because the project was neither feasible nor cost-effective. At that point, we started looking at low-cost automation alternatives, and we found out that the Karakuri concept was our solution.

KARAKURI

Karakuri means gimmick, mechanism, machinery, trick, contrivance, or device. The key point here is some mechanical trickery. It originated with mechanical dolls in Japan, called Karakuri ningyo. These dolls were first mentioned around 1500 years ago but were most popular around 200 years ago (Fig. **2**).

Fig. (2). Karakuri Dolls.

These dolls are the precursor of robots. One of the most well-known examples is the tea-carrying doll. The weight of a bowl of tea put on the tray made the doll move forward to a set distance while moving its feet (powered by a wound-up spring). After removing the bowl and drinking the tea, the empty bowl was placed on the tray again. This weight causes the doll to turn around and return to its original position.

When it comes to lean manufacturing, Karakuri is also known as Karakuri Kaizen. This mechanical device has the aim of improving processes and conveyance systems. The device uses only mechanical gadgetry and shuns electric, hydraulic, or pneumatic power. It is also not controlled by a computer but rather by the design of the mechanics.

Power

A Karakuri device needs the power to function, or more generally, it needs energy. Rather than using a dedicated power source like a motor, Karakuri devices take their energy from wherever they can. Often, the energy source is human muscle. Many Karakuri devices are operated by hand, like a lever or a pair of custom pliers. This can also be in an indirect form (*i.e.*, when the worker takes a power tool out of a holder or returns it, the movement of the power tool can power a mechanism). Many other Karakuri devices are operated by stepping on a lever or pedal. This provides more power as the muscles in the leg are stronger. Another energy source is taking away a bit of energy from another machine. The movement of another machine is used to power the K device. For example, a material supply cart driving by a storage rack may activate some levers within the rack.

Energy Storage

For many Karakuri devices, energy must be stored. Seesaw is a very common type of these devices. While one thing moves down (releases gravitational energy), the other thing moves up (stores gravitational energy). Later this stored gravitational energy can be used when the first thing has left the device.

Closely related is weight on strings. During a movement, this weight is pulled up. Later, this weight is released again to provide energy. Often, these weights are plastic bottles or canisters filled with water or sand, which allow easy fine-tuning of the weights. Yet another ingenious way of energy is a pendulum.

One example is the extraction of screws using a magnetic pendulum. This pendulum pulled the screws out of a storage container, oriented them and placed them on a small stop for the worker to use. The pendulum stored enough energy to get five to six screws, which was all that worker needed.

Principles of Movement

There are four types of movements that convert the motion of one form into other forms. Besides changing direction, this means converting from one side to another rotational motion, linear motion, rolling, and intermittent rotation (Fig. **3**).

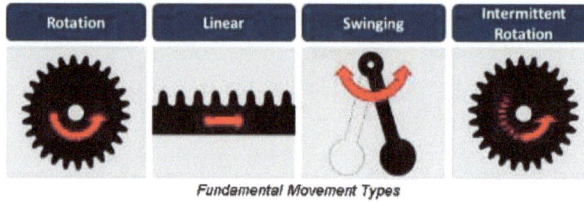

Fundamental Movement Types

Fig. (3). Different Movement Types.

There will be all kinds of gears, ropes, pulleys, cams, cogs, bars, links, belts, and whatnot. Those elements are the basic components of our Karakuris.

Implementation

As regards our main problem that has been mentioned before, after reviewing the concepts learned from Karakuri, our goal was to reduce the non-value-added activities of the operators through simple mechanisms.

To get an overview we built a Yamazumi chart of the assembly line.Yamazumi is a Japanese word that means to stack up. A Yamazumi chart (or Yamazumi table) is a stacked bar chart that shows the origin of the cycle time in a given process and it is also used to graphically represent processes for optimization purposes(Fig. **4**).

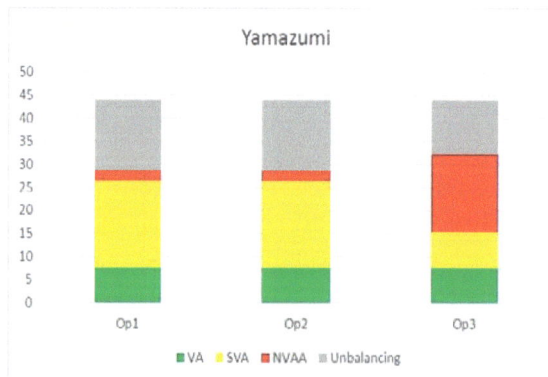

Fig. (4). Yamazumi Chart.

Process tasks are individually represented in a stacked bar chart, which can be classified as Value-Added, Semi-Value-Added, Non-Value-Added or Unbalancing [1].

Value-Added-Activities are tasks that produce a physical or chemical change in the product.

Non-Value-Added activities are those that do not produce a change in the product and the customer is not willing to pay for them, such as walking, transporting, or searching.

Semi-Value-Added-Activities are Non-Value-Added-Activities that are very difficult to eliminate, such as picking a part.

The Unbalancing is the waiting time in which the operator has nothing to do.

The average duration of each task is recorded and displayed in the bar chart. Each task in the process is stacked to represent the complete process step.

For our case, consider positioning the product in the box and placing the box on the pallet as a Value-Added-Activity. Then, grouping the product and assembling the box as a semi-value-added activity. Finally, walking and moving activities as a Non-Value-Added activity.

As we can see in the chart, operator 3 has the highest rate of Non-Value-Added activities.

To examine in depth the main non-value-added activities we use another tool: the spaghetti chart.

A spaghetti chart is a quick and easy way to keep track of the distances of parts and people in the shop. It gets its name from the fact that the result looks like a plate of spaghetti. A spaghetti diagram, also known as a spaghetti chart, spaghetti model or spaghetti graph, is a tool for determining the distance traveled by (usually) man or (in some cases) material (Fig. **5**).

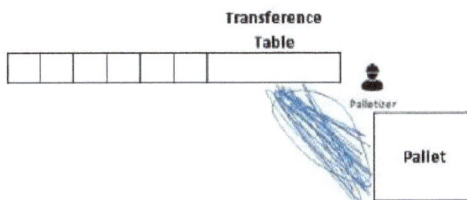

Fig. (5). Spaghetti Chart.

In these cases, it is now also possible to use the Application of Graph Theory in Workplace Design, presented by Eng. Gonzalo Taboada.

If you want to reduce the distance that parts or people travel in the workstation, a spaghetti diagram can help you achieve this goal.

Let's use it in our project.

In the workstation we can see that we have a buffer full of boxes. The third operator needs to walk from the point where he or she puts the box on the pallet to this buffer to pick up the boxes. Half of his Non-Value-Added activities are related to this walk (Fig. **6**).

Fig. (6). Workstation.

To avoid this walking process, we need the palletizer to stay next to the pallet and the boxes need to travel along this linear movement without intervention. This linear movement can be achieved with an inclined plane (Fig. **7**).

Fig. (7). Proposal Solution.

As a result, this simple and useful solution allowed us to reduce 50% of the operator's Non-Value-Added activities.

Our second challenge was to address the problem of positioning and grouping our products for insertion into the box. In order to do this, it was necessary to know that after the product was filled, it traveled on the conveyor belt in a vertical position and it must enter the box in a horizontal position(Fig. **8**).

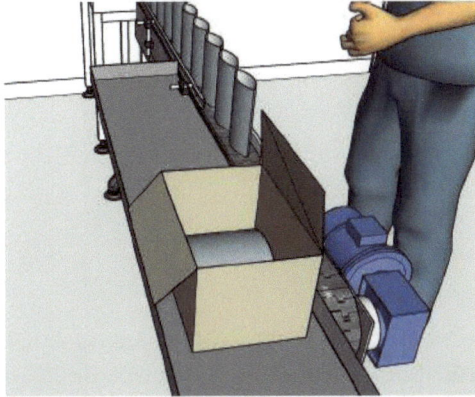

Fig. (8). Manual Solution.

In this case we decided to take advantage of the movement of the conveyor to apply a force at the top of the product so that it rotated and fell into the box. To conduct this, we added a guide that pushed the product as it moved along the belt. We also had to modify the height of the platform where we placed the box once it was assembled.

As we can see in the following figure, we eliminated the task of placing the product in the box. Now, the operators must only assemble the box and place it on the platform (Fig. **9**).

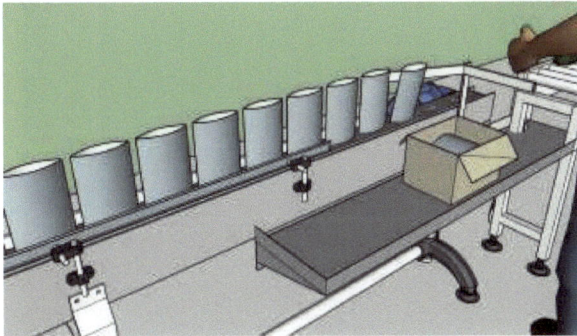

Fig. (9). Low Cost Automation.

After these two simple solutions using the low-cost automation concept and Karakuri, our Yamazumi chart looks like this (Fig. **10**):

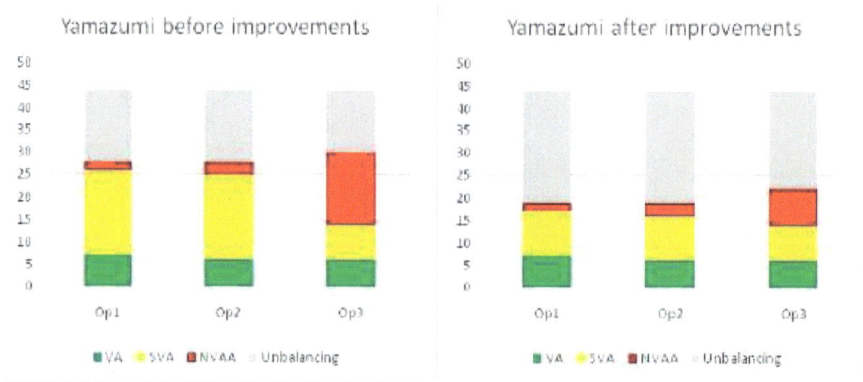

Fig. (10). Yamazumi Comparison Chart.

As we can see, operator 1 and operator 2 have a saturation level less than 50%, which allows us to perform both tasks with only one operator.

RESULTS

With these two simple solutions we were able to reduce from three operators to one and a half. This means that we use one operator to pack the products and the other operator can palletize the boxes from two assembly lines. Our Yamazumi chart after balancing looks like this (Fig. **11**):

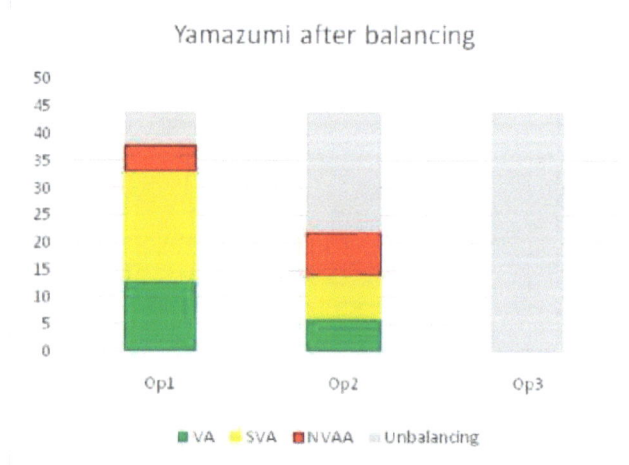

Fig. (11). Yamazumi Chart, After Balancing.

We can now assess the results in terms of productivity.

Productivity is the effectiveness of an effort, in the industry it could be measured as the rate of output per unit of input (Fig. **12**).

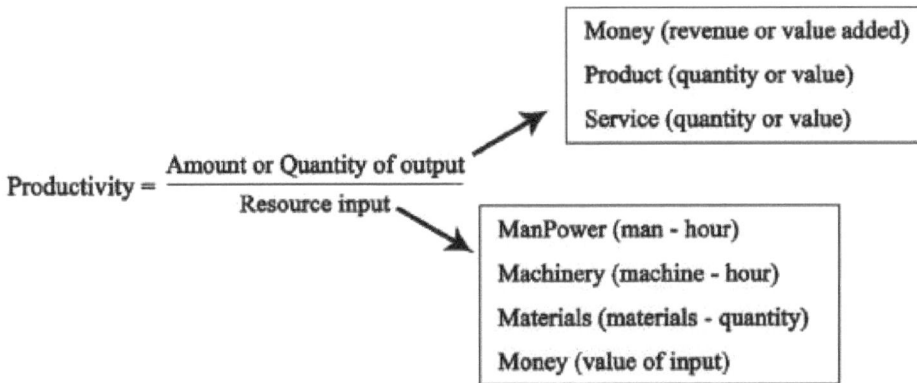

$$\text{Productivity} = \frac{\text{Amount or Quantity of output}}{\text{Resource input}}$$

Money (revenue or value added)

Product (quantity or value)

Service (quantity or value)

ManPower (man - hour)

Machinery (machine - hour)

Materials (materials - quantity)

Money (value of input)

Fig. (12). Productivity Formula.

In our case, the output is the quantity of product, which does not present any variation. The input is the manpower (man-hour). The outcome of the implemented solution was the reduction from 3 operators to 1.5. This means that we doubled productivity [2].

CONCLUDING REMARKS

The original automation proposal was to buy 3 machines, one to prepare the box to receive the product, a robot to take the product and place it into the box and another robot to palletize the box. With this solution we would have reduced from 3 operators to 1. We need to consider that this solution also increases the running costs of electricity and maintenance of the new machines.

After applying low cost automation, we were able to reduce from 3 operators to 1.5 without increasing running cost and with a very low investment.

All labor figures must be multiplied by 3, because the factory works 3 shifts per day.

In this case it is interesting to calculate the amortization period.

As we know, the payback period is the time required for the cash inflows generated by a project to offset its initial cash outflow (Fig. **13**).

$$\text{Payback Period Formula} = \frac{\text{Initial Investment Made}}{\text{Net Annual Cash Inflow}}$$

Fig. (13). Payback Formula.

Using this formula in our case:

Table 1. Payback.

	Full automation		Karakuri	
Investment	USD	350.000,00	USD	5.000,00
Annual saving	USD	60.000,00	USD	45.000,00
Annual Running Cost Increase	USD	1.500,00	USD	-

Payback period	5,98 years	0,11 years

Companies usually may establish 3 years as the maximum threshold which all projects that have been accepted must remain; therefore the full automation solution was not attainable. That is why our Karakuri solution arises to make a big difference: almost 6 years payback period *versus* less than two months.

CONSENT FOR PUBLICATION

Not applicable.

CONFLICT OF INTEREST

The author declares no conflict of interest, financial or otherwise.

ACKNOWLEDGEMENTS

Declared none.

REFERENCES

[1] S. Shingo, "A Study of the Toyota Production System", Boca Raton, FL, CRC Press Taylor & Fancis Group, 2005.

[2] B.W Niebel, and A. Freivalds, "Methods, Standards and Work Design", 11th ed. DF, Mexico: Alfaomega, 2004.

CHAPTER 6

Agile Methodologies in Manufacturing Projects

Gerardo E. Rodríguez[1,*]

[1] *Department of Industrial Engineering, Universidad Tecnológica Nacional FRC, Córdoba, Argentina*

Abstract: This article analyzes the feasibility of using agile methodologies tools in industrial manufacturing projects.

Since their inception, agile management tools have been used for software development projects and technology innovation. Currently, manufacturing projects employ only traditional project management methodologies; the challenge is to apply agile tools in traditional management.

Keywords: Agile, GPD, Kanban, Lean, Management, Methodology, PDCA, Project, Scrum, Sprint, TPS, Tools, WIP.

INTRODUCTION

Before going into the development of the article, we will briefly review the origins of agile management methodologies.

It is difficult to identify a creator of agile methodologies; in fact, there is no date in history that we can mark as the birth of agile methodologies. This is because agile methodologies are an incremental philosophy of knowledge and experience that haves been in constant evolution for more than a hundred years.

However, to understand the birth of agile tools, we must travel back in time to the year 1900` and analyze different technological milestones in history. Below we will review these milestones, and briefly explain the contribution they made to the agile methodologies we know today.

TEMPORARY REVIEW

To study the beginning of agile methodologies, it is necessary to review several contributions throughout history, starting in 1891, Frederick Taylor (1856-1915),

* **Corresponding author Gerardo E. Rodriguez:** Department of Industrial Engineering, Universidad Tecnológica Nacional FRC., Córdoba, Argentina; Tel: +5493516716992; E-mail:ger_086@live.com.ar

who first defined the concepts of the Scientific Administration of Labor. He extended his ideas on division of labor to simple and routine tasks, motivation and productivity through control and supervision of employees. (USA-1891).

In 1924, Sakichi Toyoda (1867-1930) in Japan created and implemented process automation for the first time with the help of humans, concept known as *Jidoka*. Invent the loom machine capable of automatically detecting when the thread breaks and stopping production from avoiding waste or defects in the final fabric. Later at Toyota Motor Company, Sakichi Toyoda would apply these concepts known today as Lean Manufacturing [1] (Fig. **1**).

Fig. (1). First Toyoda Loom.

In 1936 Alan Turing (1912-1954 England) developed the Turing Machine, which explained years later the logical operation of a computer. It was the first machine to use symbol processing using logic (Fig. **2**).

Fig. (2). Turing machine.

Walter Shewhart (1891-1967 USA), known as "The father of quality", in 1939 published in his book Statistical Method from the Viewpoint of Quality Control the process of continuous improvement and iterative and incremental work through short cycles of " Plan, Do, Check and Act "(plan-do-study-act). "P.D.C.A."

In 1946, in the USA, the ENIAC was created, the first digital computer in history. ENIAC is the acronym for "Electronic Numerical Integrator and Computer" It was the first general-purpose computer, with the possibility of being reprogrammed to solve numerical problems. It was initially designed to calculate artillery firing charts destined for United States Army military research.

In 1948 the engineer Taiichi Ohno (1912-1990) in Japan, begins to create the Kanban methodology in Toyota, to give rise to what is known as "Toyota Production Systems" (TPS).

In 1950, in the USA, the X-15 hypersonic jet was developed. It was the first technology project developed with "Iterative and Incremental methodology" recorded in history (Fig. **3**).

Fig. (3). X-15 jet.

NASA in 1958 used an Iterative and Incremental methodology for the Mercury project. (for the first trip of a man to space). Incremental methodology begins to have more military uses and in research developments.

Between 1966 and 1969, the United States Department of Defense developed the first computer network, which was called ARPANET, considered this milestone as the birth of the Internet.

In 1970 Dr. Winston Royce (1925-1999) wrote the paper that made history "Managing the Development of Large Software Systems". This paper has been attributed many times the first definition of the sequential process for software development. Today known as "Waterfall". But, if we take a closer look at the Paper, Royce does not recommend this method, he anticipates its risks and suggests that it should be iterated at least twice. In the paper he mentions that the waterfall process "invites failure" (Fig. **4**).

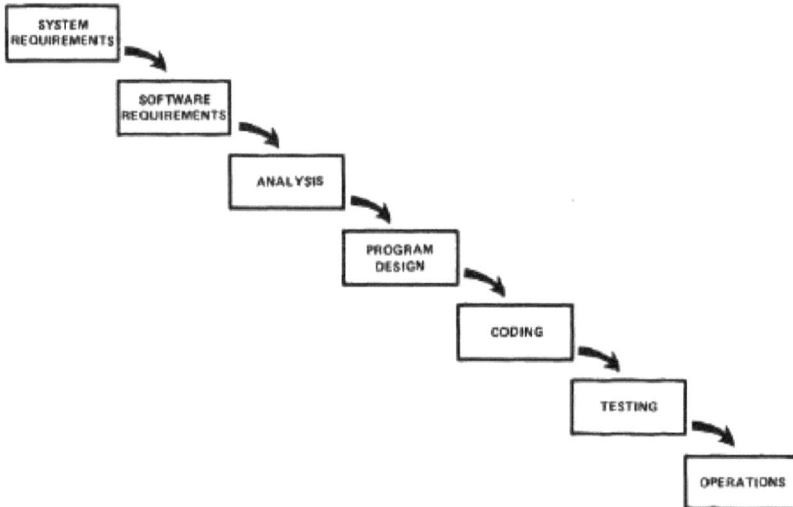

Fig. (4). Waterfall process – Steps to develop a computer program.

The next three milestones in history, for many authors, are the milestones responsible for all the technological advances that we know and use today. All the technology available today has to do with the milestones these:

- 1974: The term Internet is used for the first time to refer to the network (originated as ARPANET)
- 1975: Bill Gates and Paul Allen found Microsoft ®
- 1976: Steve Jobs and Steve Wozniak create Apple Computers Inc ®.

Starting in 1976, software and technology companies used cascade methodology (waterfall) for the development of their projects. But for many of these companies, the methodology was not successful, since they were unable to complete their projects in a timely manner.

Competition between companies, forced to be more agile and faster in innovation to meet the needs of customers.

Technology companies needed to find the most agile, and most accurate way to work on product development. They needed to have integration between work teams and be able to achieve faster results.

For these reasons, we begin to question the cascade methodology, and we begin to think of more agile solutions. Work begins with different tools and is used as an incremental iteration as the basis for each stage of a project.

In 1986, the first use of the term scrum for product development is recorded in the article "The New New Product Development Game" published in the Harvard Business Review. The authors, Takeuchi Hirotaka and Nonaka Ikujiro applied these methodologies in companies such as Fuji-Xerox, Canon, Honda, NEC, Epson, Brother, 3M, Xerox, and Hewlett-Packard.

This tool is considered the first agile management tool for project development. In reality, it is a team management tool, where you work collaboratively and incrementally.

The term "Agile" is commonly related to the term "Lean", and although they are not the same terms, both have the reduction of tasks that do not add value and be faster in making decisions in a production process.

It is important to note that the term Lean had its origins in 1988, when the term Lean Manufacturing was first used in the article "Triumph of the Lean Production System" (John Krafcik).

Later, in the year 1990, the term Lean Production re-referred to Toyota Production System is popularized in the book "The Machine That Changed the World" (James Womack).

From 1990, the technology companies adopted agile management tools, leaving behind the rigid methodologies based on the waterfall process.

In 1994, Standish Group International Inc. (an independent international IT research advisory company) publishes a study known as the "CHAOS Report" which shows the low results of Software Development projects, where "The Waterfall" being the standard model here.

- 52.7% of projects exceeded costs and/or time
- 31.1% of projects failed, were canceled
- 16.2% of projects were successful

These results confirm that the waterfall methodology was not successful for technological projects. With which it was necessary to change the way of developing the projects.

The next milestone in history can be considered as the formal birth of agile methodologies. In 2001, 17 industry professionals got together to discuss and sign what is known as the "Agile Manifesto".

This manifesto, promoted by Kent Beck (American software engineer, one of the creators of extreme programming software development methodologies (eXtreme Programming or XP), who currently works for Facebook, is considered one of the creators of the agile methodologies that we know today.

It is important to note that in 2001, the book "The Toyota Way" with the 14 Lean Manufacturing Principles was published. This allowed the publication of the book "Lean Software Development" (Mary and Tom Poppendieck) in 2003, bringing the Principles of Lean manufacturing to Software development.

SCRUM & KANBAN - "THE AGILE TOOLS"

Previously we reviewed the history of project development and the birth of agile methodologies.

The Agile methodologies are a set of tools, in this article we are going to develop the Scrum & Kanban tools, which is considered the most powerful and currently used tool for project development.

Before developing these tools, it is important to compare traditional and agile methodologies:

Manufacturing Projects (Traditional)	Agile projects
Predictive	Adaptive
Process-oriented	People-oriented
Rigid process	Flexible process
It is conceived as a project	Project subdivided in several other project's little ones
Little communication with the client (in/out)	Constant communication with the client
Final product delivery to finish development	Constant deliveries of partial product
Extensive documentation	Little documentation

As we can see from the comparison table, a traditional manufacturing project in general is very different from an agile project. The main differences to consider are the following:

- A traditional manufacturing project has no contact with the customer, until at least it has a finished product (pre-series).
- A traditional manufacturing project does not have partial project deliveries, it has stages, but everything is considered as a single project.

It is important to understand that a project with agile methodologies is a continuous process of iterations. In each iteration we look for a "deliverable" that is of value to the client and that allows us to move on to the next stage, but validating the work already done.

In the following pages, we will first develop scrum as the main teamwork tool and then Kanban as a complementary tool.

Scrum

Its name does not correspond to an acronym, but to a sporting concept, specific to rugby, related to the training necessary for the rapid recovery of the game in the event of a minor foul. Its first reference in the context of development dates back to 1986, when Takeuchi and Nonaka used the Rugby Approach to define a new approach to product development, aimed at increasing its flexibility and speed, based on the integration of an interdisciplinary team and multiple overlapping phases.

The Scrum methodology is a framework designed to achieve effective team collaboration in projects. It uses a set of rules that define the roles that generate the necessary structure for its correct operation.

Scrum uses an incremental approach based on empirical process control theory. This theory is based on:

- Transparency
- Inspection
- Adaptation

Transparency, which ensures visibility into the process of things that can affect the outcome. Inspection, which helps to detect undesirable variations in the process. Adaptation, which makes appropriate adjustments to minimize the impact of variations.

Scrum Teams" are self-managed, cross-functional and work in iterations. Self-management allows you to choose the best way to do the work, rather than having to follow guidelines from people outside the team.

Team members have all the necessary knowledge to carry out the work, and product delivery is done in iterations, with each iteration creating new functionality or modifying functionality required by the product owner.

Scrum defines three roles:

- the Scrum masters
- the product owner
- the development teams.

The Scrum master's role is to ensure that the team is adopting the methodology, its practices, values, and standards; he/she is the team leader but does not manage the development.

The product owner is a single person and represents the stakeholders; he is responsible for maximizing the value of the product and the work of the development team; his functions include managing the ordered list of required functionalities or Product Backlog.

The development team, on the other hand, is responsible for converting what the customer wants, the Product Backlog, into functional iterations of the product. The development team has no hierarchy, all its members have the same level and position: "developers".

The optimal team size is between three and nine people, but there is no rule for this.

Scrum defines a main event or Sprint (Fig. **5**) that corresponds to a time window where a usable version of the product (with increased value). Each Sprint is considered an independent project, and the maximum duration is one month (recommended).

A sprint it consists of the following elements:

- Sprint Planning Meeting
- Daily Scrum Meeting
- Sprint Review Meeting
- Sprint Retrospective Meeting

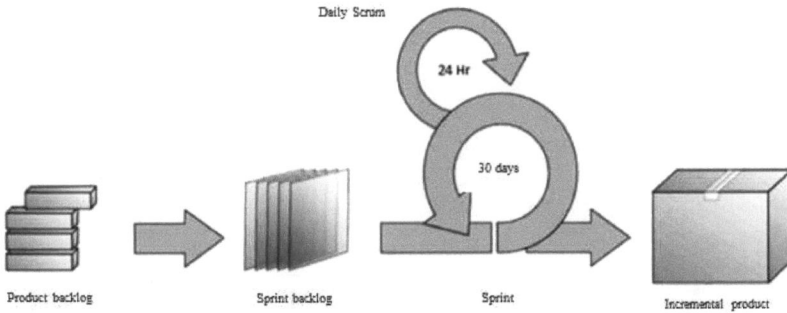

Fig. (5). SCRUM / SPRINT phases.

The Sprint Planning Meeting defines the work plan: what is to be delivered and how it will be achieved. It is the design of the system and the estimation of the amount of work.

It is usual for this activity to last eight hours for a one-month Sprint, if the Sprint has a shorter duration, the time is allocated proportionally.

The daily Scrum is a development team event of fifteen minutes maximum. It is held every day to explain:

• what has been achieved since the last meeting
• what will be done before the next meeting
• and the obstacles that have arisen

This event is developed through a meeting that is normally held standing up with the participants gathered in a circle, to prevent the discussion from spreading out.

The review takes place at the end of the Sprint and lasts four hours for a one-month Sprint (or a proportion of that time if the duration is shorter).

In this phase, the project owner reviews what has been done, identifies what has not been done and discusses the Product Backlog. The team explains the problems encountered and how they have been solved and shows the product and how it works. This meeting is of great importance for the following Sprints (Fig. **5**).

The Sprint Retrospective is a Scrum Team meeting where they discuss:

• How was the communication
• The process and tools
• What was good, what was not

The Product Backlog is a list (sorted by value, risk, priority and need) of requirements that the product owner defines, updates and orders.

The list has the characteristic that it is never finished, it evolves during the development of the project.

The Sprint Backlog is a subset of elements in the Product Backlog and the plan to carry out in the product increment.

As the Product backlog is organized by priorities, the Sprint backlog is built with the highest priority requirements from the Product backlog and with those that were left unresolved in the previous Sprint.

Once built, the Sprint backlog must be accepted by the development team.

Progress tracking consists of the sum of the work remaining to be done in the Sprint. It has the characteristic that it can be given at any time, which allows the product owner to evaluate the progress of the development.

To make this possible, team members constantly update the status of requirements they have assigned indicating how much they consider that they to finish.

Kanban

We previously saw that Kanban was born in the 1950s, and over time it became one of the most widely used tools in the automotive industry.

Kanban is a method of managing work Toyota Production System (TPS). It was implemented to be able to work with the "just in time" system that represents a drag system. This means that production is based on the demand of the customers and not in the traditional "pull" practice of making products and trying to sell them in the market.

Its main objective is to minimize waste without affecting production. The main objective is to create more value for the customer without generating more costs.

What Does Kanban Mean?

The word Kanban comes from Japanese and literally translated means "card with signs" or "visual signal". The most basic Kanban board is made up of three columns: "To do", "In process" and "Done". If applied well and working correctly, it would serve as a source of information, as it demonstrates where the bottlenecks are in the process and what is preventing the workflow from being continuous and uninterrupted (Fig. **6**).

Fig. (6). KANBAN BOARD (Toyota 1950).

In the early 21st century, the software industry discovered that kanban could make a real difference in the way products and services were produced and delivered.

Kanban proved to be suitable not only for the automotive industry, but also for any other type of industry. Thus, the Kanban method for agile management methodologies was born (Fig. 7).

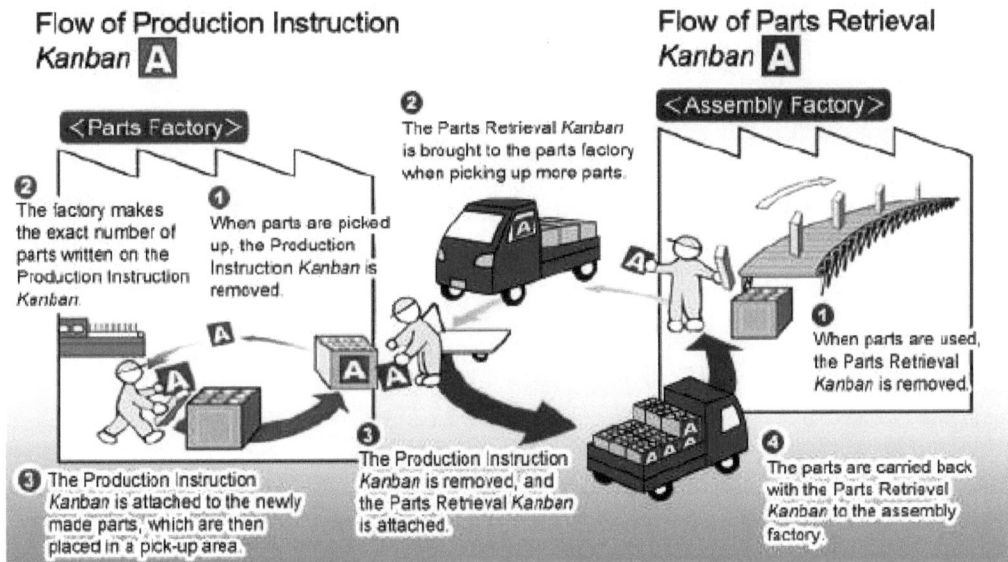

Fig. (7). KANBAN FLOW (all right reserved TOYOTA company).

Kanban Process

Steps

1- Visualize the work in Kanban and the phases of the production cycle, or workflow.
2- Determine the limit of "work in progress".
3- Measure time to complete a task.

In this last step, it is important to determine the time it takes to complete each task to be measured. This time is called the "lead time". The lead time counts from the time a request is made until delivery is made.

Although the most well-known Kanban metric is "lead time", another important metric is also often used: "cycle time".

Cycle time measures the time from the start of work on a task until it is completed.

If "lead time" measures what customers see, what they expect; "cycle time" measures more the performance of the process.

Final consideration on Kanban and Scrum

Nowadays it is common to use both tools at the same time, but there are important differences between the two methods:

- Kanban's rules are far less than Scrum's.
- Kanban does not define iterations (Sprints).
- Kanban explicitly limits the tasks that can be performed per phase (with the work in progress limit).
- Scrum limits work indirectly through sprint planning.

Before proceeding, it is important to ask the following questions:

- What are the most important variables in a project?
- What are the factors that help the success of a project?
- What are the factors that can make a project fail?

Next, we will propose and evaluate the feasibility of using agile tools in heavy manufacturing projects.

AGILE TOOLS IN MANUFACTURING PROJECTS

There are different project management methods; currently, each company develops its projects following its own management method.

It is important to know what these methods have in common in order to understand how they work.

We can take as a reference the project management following the traditional methodology proposed by the PMI (Project Management Institute), but there are other methods. For this article we are going to use the GPD (Global Product Development) method used by the automotive industry.

Manufacturing projects in general have the following characteristics:

- Extensive time
- Large work team
- People from different places
- Rigid deadlines
- Minimal or no contact with the end user

So, with these characteristics, is it feasible to use agile tools? The answer is YES, and this is because the tools are oriented to the management of teams and tasks, and not to the product itself.

The challenge is to make the team work quickly!

This is a typical automotive product development process (Fig. **8**):

Fig. (8). GPD (global product development) process.

The process is divided into deliverables and milestones, each of which has a validation stage and an estimated time.

It is important to note that this method requires a lot of planning and development time and we only have product development in the advanced stages of the project.

The end customer does not participate in most of the project. And only in the advanced phases of development (Pre-build) and (4P) do we have a tangible product that can be validated.

This is very different from a project with an agile methodology, where at each stage of the project we have a physical deliverable to be validated, and with the participation of the end user.

The GPD has as a tool to center a project calendar (schedule), which is directly related to the company's requirements and will attend to the manufacturing's operation plan (MOP).

The GPD team is a cross-functional team, from different areas of the company, and the whole project is controlled by a specific area, with the support of others (product launch / validation / quality / engineering) (Fig. **9**).

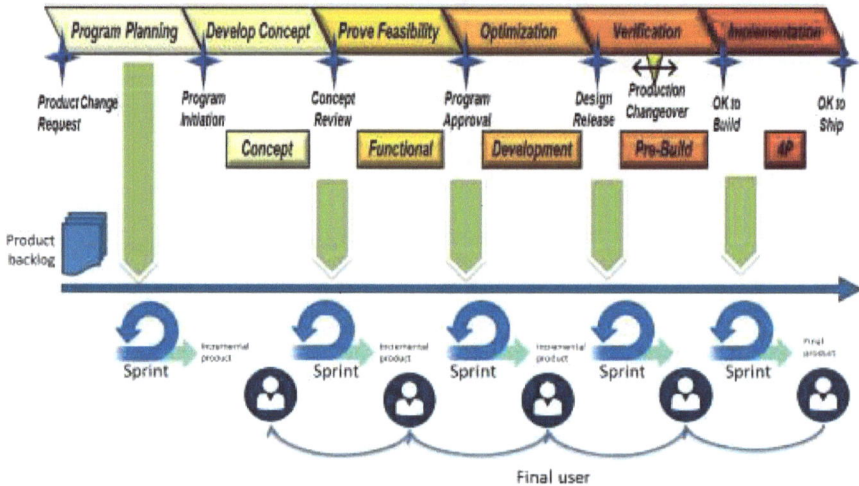

Fig. (9). GPD + SCRUM process.

GPD + SCRUM

To use scrum in a traditional project, it is important to be able to divide the project into stages and define sprints for each one. Each stage must provide added value, it must deliver an incremental product.

This is a very important point, as it is the scrum philosophy, the work must always be incremental, and the final assumption must be present during the project (to validate each progress).

The goal is that an agile tool coexists with a traditional one. The tools must complement each other and have the same objective: To fulfill the project in time and form, satisfying the client.

Depending on the timeframe and stages of the project, the team must define the correct deadline for sprinting. It will also be necessary to define the roles and how each member responds to the different areas of the company.

If we can combine scrum tools with the traditional model, it will also be possible to follow the project and manage it through Kanban.

GPD + SCRUM + KANBAN

The main function of Kanban is to regulate WIP (work in process/progress). This concept in a traditional project is often not managed, and there are problems in assigning tasks to team members.

It is common for people to feel that they do not have the time needed to perform the tasks of a stage, and this is due to misallocation of time and overloading people with work.

GPD + SCRUM + KANBAN

In manufacturing plants, the concept of WIP is very common, and refers to the control of work-in-process to increase productivity. Work-in-process control is a necessity for world-class companies to increase competitiveness (Fig. **10**).

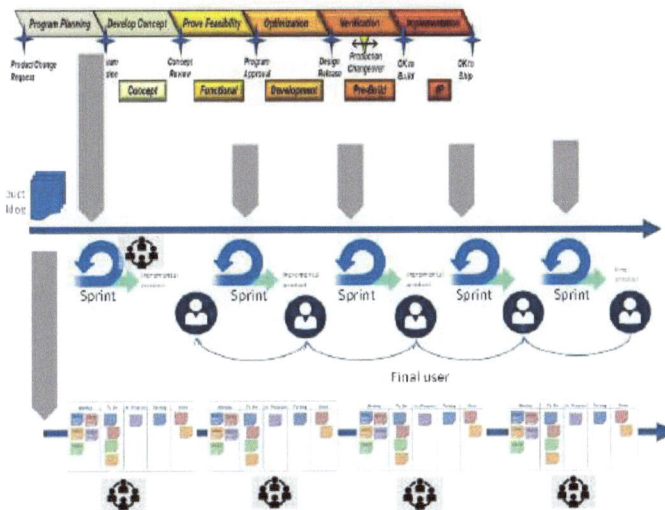

Fig. (**10**). GPD + SCRUM + KANBAN process.

How do we include Kanban in a traditional project?

In the planning stage of the project, activities are initiated, which we call the backlog. These tasks are placed on a Kanban board, and the team as a whole must monitor the progress of the tasks.

It is important that in a Kanban tool, everything is visual, management is available to everyone all the time. This way it is easy to see if a task is delayed and to help.

The Kanban board can be manual or electronic. Currently there are different digital tools, but it is advisable to have a physical board for sprint reviews. An electronic board helps team communication when people are not physically in the same location (Figs. **11** and **12**).

Fig. (11). Manual KANBAN board example.

Example - KANBAN project management board:

Fig. (12). Electronic KANBAN board example.

CONCLUDING REMARKS

Agile methodologies work well in a specific context characterized by small, co-located development teams with customers who can make decisions about requirements and their evolution.

With requirements that change frequently (weekly, monthly), with variable project scope or budget, and with few constraints in the development process.

Outside this environment, it is common for problems to arise due to lack of client involvement, in architecture, design intensive or documentation intensive projects; there is a slow pace of change and distributed teams.

The role of the customer is vital in the development process of agile methodologies, remember that one of the principles of the "Agile Manifesto" states that business managers and developers work together on a daily basis throughout the project.

This condition is not easy to satisfy. Customers may be geographically separated, or it may be too costly for them to maintain a representative with the capacity to respond to all the requirements of the system being developed, on a permanent basis.

Several strategies can be used to address these limitations:

- Have multiple user story owners
- Have a client intermediary

Another difficulty, regardless of the context, has to do with the focus of the documentation on the evolution and maintenance of the product. It derives from the concept issued by agile experts, who indicate that documentation should be brief, precise and limited to the fundamental functions of the system, because having the product running is the most important thing.

The view that system quality is achieved through iterative development and not through documentation leads to long-term problems, since documentation is the medium that helps knowledge transfer and product maintenance and evolution.

So, again: Is it feasible to use agile tools in manufacturing project? → YES, but:

- The project should be divided into small stages
- Each stage should result in an incremental value product
- Each stage must have a small responsible team (maximum 20 people)

- Work teams must be multidisciplinary
- Project members must work exclusively for the Project
- The end customer or end user must participate in the early stages of the Project
- Each project progress must be validated by the manager, and with the client's point of view.

To conclude, we will compare a rigid project, a hybrid project, and an agile project.

Today, many companies say they are agile in their projects, but in reality, they are managed in a mixed way. They are not 100% agile!

Rigid project

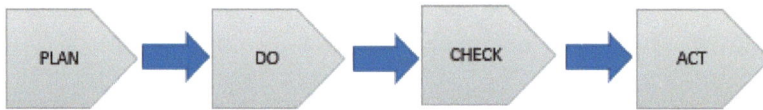

Mix project

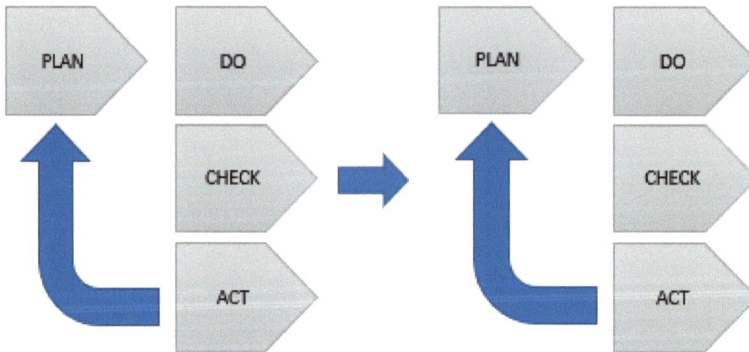

Agile project

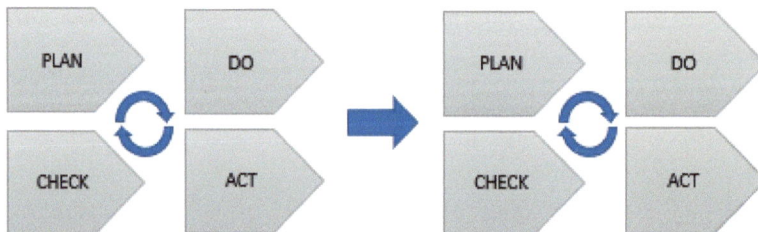

CONSENT FOR PUBLICATION

Not applicable.

CONFLICT OF INTEREST

The author declares no conflict of interest, financial or otherwise.

ACKNOWLEDGEMENTS

Declared none.

REFERENCES

[1]　　C. Larman, "Cover feature", *IIEE Computer Society,* vol. 03, pp. 47-56, 2012.

[2]　　J. Womack, D. Jones, and D. Roos, "The Machine That Changed the World", In: *New York: Simon & Shuster, ,* 2007..

APPENDIX

Graph Theory

In mathematics, graph theory is the study of graphs, which are mathematical structures used to model pairwise relations between objects. Conceptually, a graph in this context is made up of vertices (also called nodes or points) which are connected by edges (also called links or bridges). A distinction is made between undirected graphs, where edges link two vertices symmetrically, and directed graphs, where edges link two vertices asymmetrically; see Graph (discrete mathematics) for more detailed definitions and for other variations in the types of graph that are commonly considered. Graphs are one of the prime objects of study in discrete mathematics (Fig. **1**).

The following are some of the more basic ways of defining graphs and related mathematical structures.

In one restricted but very common sense of the term, a graph is an ordered pair $G = (V, E, \phi)$ comprising:

V, a set of vertices (also called nodes or points)

$E \subseteq \{\{x, y\} \mid x, y \in V \text{ and } x \neq y\}$, a set of edges (also called links or bridges), which are unordered pairs of vertices (that is, an edge is associated with two distinct vertices).

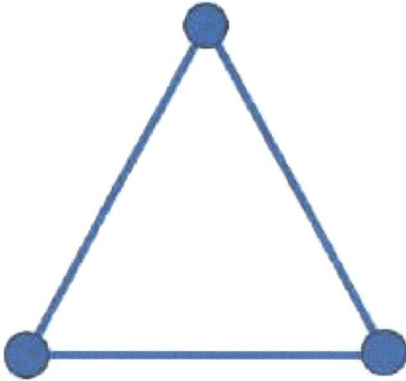

Fig. (1). Graph with three vertices and three edges.

This type of object may be called precisely an undirected simple graph.

In the edge $\{x, y\}$, the vertices x and y are called the endpoints of the edge. The edge is said to join x and y and to be incident on x and on y. A vertex may exist in a graph and not belong to an edge. Multiple edges, not allowed under the definition above, are two or more edges that join the same two vertices.

In one more general sense of the term allowing multiple edges, a graph is an ordered triple G = (V, E) comprising:

V, a set of vertices (also called nodes or points)

E, a set of edges (also called links or bridges)

ϕ: E $\subseteq$ {{x, y} | x, y $\in$ V and x $\neq$ y}, an incidence function mapping every edge to an unordered pair of vertices (that is, an edge is associated with two distinct vertices).

This type of object may be called precisely an undirected multigraph.

A loop is an edge that joins a vertex to itself. Graphs as defined in the two definitions above cannot have loops, because a loop joining a vertex x to itself is the edge (for an undirected simple graph) or is incident on (for an undirected multigraph) {x, x} = {x} which is not in {{x, y} | x, y $\in$ V and x $\neq$ y}. So to allow loops the definitions must be expanded. For undirected simple graphs, the definition of E should be modified to E $\subseteq$ {{x, y} | x, y $\in$ V}. For undirected multigraphs, the definition of ϕ should be modified to ϕ: E $\rightarrow$ {{x, y} | x, y $\in$ V}.

These types of objects may be called undirected simple graph permitting loops and undirected multigraph permitting loops, respectively.

V and E are usually taken to be finite, and many of the well-known results are not true (or are rather different) for infinite graphs because many of the arguments fail in the infinite case. Moreover, V is often assumed to be non-empty, but E can be the empty set. The order of a graph is |V|, its number of vertices. The size of a graph is |E|, its number of edges. The degree or valency of a vertex is the number of edges that are incident to it, where a loop is counted twice.

In an undirected simple graph of order n, the maximum degree of each vertex is n − 1 and the maximum size of the graph is n (n − 1)/2.

The edges of an undirected simple graph permitting loops G induce a symmetric homogeneous relation ~ on the vertices of G that is called the adjacency relation of G. Specifically, for each edge (x, y), its endpoints x and y are said to be adjacent to one another, which is denoted x ~ y.

Glossary

Applied Design:	Design that considers factors of manufacturing, methods and the behavior of the product.
CILR:	Cleaning, Inspection, Lubrification and Refastening.
GPD:	Global Product Development, is when a company has their product development activities globally distributed, from R&D to production.
Graph:	Graphical representation for solutions of discrete mathematics.
JIT:	Just in Time, is a methodology aimed primarily at reducing times within the production system as well as response times from suppliers and to customers.
Kanban:	Is one method to achieve JIT. The system takes its name from the cards that track production within a factory. For many in the automotive sector, Kanban is known as the "Toyota nameplate system".
	Japanese mechanical automaton devices to produce surprise in a person
Lean Manufacturing:	Efficiency manufacturing process
NVAA:	Non-value-added-activities for efficiency process.
PDCA:	It is the Deming cycle, plan-do-check-act to solve problems.
CILR:	Cleaning, Inspection, Lubrification and Refastening.
Supported Employment:	Methodology based on experimental psychology and behavior modification through reinforcement, intended for pathologies with intellectual and physical disabilities.
Scrum:	Concept sport, typical of rugby, related to training required for the quick recovery of the game before a minor offense.
SWOT:	SWOT analysis is a type of diagram widely used in business and education used for exploring strengths, weaknesses, opportunities, and threats in a given situation.
TPS:	The Toyota Production System is a lean manufacturing system which entrusts team members with well-defined responsibilities to optimize quality by constantly improving processes and eliminating unnecessary waste in natural, human and corporate resources.
WCM:	World Class Manufacturing, it's the same to Lean Manufacturing.
WIP:	The term "work-in-process" to mean a manufacturer's inventory that is not yet completed, as the goods that are on the factory floor of a manufacturer
WOW System	System, without washers for banjo fitting.
Yamazumi:	A Yamazumi chart (or Yamazumi board) is a stacked bar chart that shows the source of the cycle time in a given Process. The chart is used to graphically represent processes for optimization purposes.

SUBJECT INDEX

A

Agile 79, 83, 84, 85, 90, 91, 92, 95
  management tools 79
  methodologies 79, 84, 85, 92, 95
  methodologies tools 79
  methodologies work 95
  solutions 83
  tools 79, 90, 91, 92, 95
Aging, active 44
Aluminum washer 26
Assembly 18, 27, 33, 39
  performance 33
  process 18, 27
Assessment 38, 39
  functional capability 38
Autism spectrum disorder 40
Automate processes 69
Automation 38, 69, 70, 78, 80
  implemented process 80
  projects 69
  solution 78
Autonomous maintenance works 11

B

Banjo 24, 26, 27, 28
  connector 26, 27
  fasteners 27
  fitting 24, 28
  screw 24
Bridges problem 6
Budget staffing 57

C

Cartography 4
Cash 77
  inflows 77
  outflow, initial 77
CILR Route 13

Cognitive

Cognitive categories 42
Computer 4, 82
  network design 4
  program 82
Copper hardens 24
Customer 48, 53, 54
  compliance 48, 53, 54
  service 48, 53, 54
Cyber-physical systems 38

D

Data analytics 38
Design, workstation 7
Development 40, 44, 82, 83, 84, 95
  software 82, 83, 84
  sustainable 44
  process 95
  services 40
Devices 5, 15, 37, 70, 71
  mechanical 71
  mounting 5
  telemetry 15
Digital processes 38

E

Efficient management system 55
Employees, strategic planning influence 47
Employment 35, 39, 43
  competitive 39, 43
Energy 71, 72
  source 71
  storage 71
Environment 20, 36, 37, 38, 40, 43, 47, 49, 50, 57, 59, 60, 64, 65, 67
  company's 50
  inaccessible 36
  urban 38
  working 20
Euler's theory 3

M

Machine learning 44
Manufacturing 18, 20, 71, 80, 84
  efficient 18
  lean 20, 71, 80, 84
Manufacturing projects 79, 84, 85, 90, 91, 95
  heavy 90
  industrial 79
  traditional 84, 85
Manufacturing's operation plan (MOP) 92
Matching strategy 51
Mathematical 1, 2, 4
  analysis 1
  modeling 4
  research 2
Mechanical 24, 70, 71
  gadgetry 71
  solution 24
  trickery 70
Mercury project 81
Methodologies 20, 39, 40, 44, 45, 47, 48, 79,
     82, 83, 86, 89, 91
  agile management 79, 89
  cascade 82, 83
  modern 47
  traditional 91
  traditional project management 79

N

Non-value-adding activities (NVAA) 18, 27

O

Organizational 36, 44, 48, 51, 52, 58, 63
  challenges 48
  commitment 63
  culture 36, 44, 51
  frameworks 44
  learning 58
  politics 52

P

Payback formula 78
Pendulum, magnetic 71
Personal growth and development 65
Physical fatigue 19
Plan 50, 61, 92
  manufacturing's operation 92
  organization's marketing 61
  staff training 50
  strategic communication 61
Planning 48, 55, 56
  process works 48
  systems 55
  workforce 56
Planning process 48, 55, 56
  strategic 55, 56
Power, pneumatic 71
Private organizations 44
Problems, numerical 81
Procedures, company's 66
Product 9, 20, 83, 85, 86, 87, 88, 91
  backlog 86, 87, 88
  delivery 86
  design 9, 20
  development 83, 85, 91
Programs 35, 40, 63
  flexible 40
  formal appraisal 63
  vocational training 35
Progress 37, 50, 59, 88, 90, 92, 94
  company's 50
  work in 90
  tracking 88
Projects 50, 60, 61, 68, 69, 79, 82, 83, 84, 85,
     90, 91, 92, 93, 95, 96
  calendar 92
  development 83, 84
  hybrid 96
  management methods 91
  progress 96
  software development 79
Public services 4

U

Urban planning 4

V

Value-added-activities 73
Vision, company's 68

W

Washers 7, 8, 9, 10, 18, 24, 25, 26, 27, 29, 30,
        31, 32
   copper 25, 27
   double 10
   elastic conical 29
   helical 29
Waterfall 82, 83
   methodology 83
   process 82, 83
Work 35, 38, 39, 40, 42, 43, 58, 60, 62, 63,
        67, 77, 81, 83, 85, 86, 87, 88, 90, 93, 95
   developers 95
   employee's 60
   factory 77
   incremental 81
   performance 62
   plan 87
   quantitative 58
Workplace organization 1
World 36, 69
   class manufacturing 69
   health organization (WHO) 36